Phosphate Fibers

TOPICS IN APPLIED CHEMISTRY

Series Editors: Alan R. Katritzky, FRS
Kenan Professor of Chemistry
University of Florida, Gainesville, Florida

Gebran J. Sabongi
Laboratory Manager, Encapsulation Technology Center
3-M, St. Paul, Minnesota

Current volumes in the series:

BIOCATALYSTS FOR INDUSTRY
Edited by Jonathan S. Dordick

HIGH-TECHNOLOGY APPLICATIONS OF ORGANIC COLORANTS
Peter Gregory

INFRARED ABSORBING DYES
Edited by Masaru Matsuoka

LEAD-BASED PAINT HANDBOOK
Jan W. Gooch

ORANOFLUORINE CHEMISTRY
Principles and Commercial Applications
Edited by R. E. Banks, B. E. Smart, and J. C. Tatlow

PHOSPHATE FIBERS
Edward J. Griffith

POLY (ETHYLENE GLYCOL) CHEMISTRY
Biotechnical and Biomedical Applications
Edited by J. Milton Harris

RADIATION CURING
Science and Technology
Edited by S. Peter Pappas

RESORCINOL
Its Uses and Derivatives
Hans Dressler

TARGET SITES FOR HERBICIDE ACTION
Edited by Ralph C. Kirkwood

A Continuation Order Plan is available for this series. A continuation order will bring delivery of each new volume immediately upon publication. Volumes are billed only upon actual shipment. For further information please contact the publisher.

Phosphate Fibers

Edward J. Griffith

Monsanto Company
St. Louis, Missouri

Springer Science+Business Media, LLC

Library of Congress Cataloging-in-Publication Data

On file

ISBN 978-1-4899-1047-9 ISBN 978-1-4899-1045-5 (eBook)
DOI 10.1007/978-1-4899-1045-5

Originally published by Plenum Press, New York in 1995
Softcover reprint of the hardcover 1st edition 1995

10 9 8 7 6 5 4 3 2 1

To the memory of Grete

Foreword

The importance of phosphate is well known to biochemists. This small, simple molecule of phosphorus and oxygen is at the center of the metabolic life of all creatures, from bacteria to humans. Virtually every facet of existence depends on phosphate—from deriving the energy in food to fueling the growth, locomotion, and nerve impulses of cells and organisms and on and on to being the backbone of DNA, the molecule of heredity.

What is known to very few in the scientific world is a very large molecule, a chain of many hundreds of phosphates, called polyphosphate. This polymer of phosphates likely existed on earth long before the familiar molecules of living cells, such as proteins and DNA, came into being and may even have had a share in their evolution. Just as intriguing is the current ubiquity of this polymer of phosphate in every living cell—bacterial, fungal, plant, and animal, The possible functions of polyphosphate in cells cover a wide range—from serving as a reservoir of energy to protecting cells from stress and aging.

In contrast to the mystery surrounding the functions of polyphosphate in living systems is the variety of uses to which this polymer has been put for many years by industry. A delightful account of these many uses—in toothpaste, food preservation, water supply, etc., is given here, but the burden of this book is how a truly wonderful use of polyphosphate has been forfeited as a result of the litigiousness of our American society.

Edward Griffith, an eminent industrial chemist at the Monsanto Chemical Company in St. Louis, was motivated in the early 1980s by the scare of asbestos as a carcinogen—grossly exaggerated, as he later learned—to find a safe substitute. Using his skills in phosphate chemistry, he was able to create fibers of polyphosphate, indistinguishable from asbestos in its insulating and nonflammable features, but with the benefit of being a body constituent, biodegradable, and virtually harmless. Despite the enormous potential of fibrous polyphosphate as a material for building, clothing, insulation, brake linings, and a dozen other applications the project was dropped as a result of legal advice. Although polyphosphate is utterly unlike asbestos, it was believed that it would still be regarded as a mineral fiber, and might be vulnerable to lawsuits which would inevitably be brought for any illness sustained by someone making or using the polymer.

In this account, Griffith supplies generous details of the inorganic and physical chemistry of phosphates, the variety of their polymeric forms, their crystallization, and the industrial means of fabricating fibers. To the nonchemist, he offers insights and reflections on scientific life in industry and the frustrations of fashioning a safe and useful product, only to have it abandoned for fear of specious attack and the incapacity of the public to evaluate risk and safety. In this vein, he mourns the abrupt elimination of other useful products, such as phosphate detergents, the most effective cleansing agents ever known, falsely accused of being a major pollutant.

Suffused through the book is Griffith's love of chemistry and of phosphates in particular. He is saddened by the poor public image of chemistry, exemplified in the action of the DuPont company to alter its age-old slogan "Better things for better living, through chemistry," a message that informed the public of the value of plastics, herbicides, and industrial chemicals for our individual and collective well-being. "Through chemistry" has been dropped in response to public awareness that chemicals, as is true of all things, natural or man-made, can be toxic too. Ingenious and charming in his numerous attacks, Griffith pleads for a rational and sober evaluation of what chemistry has provided and can still offer to make our lives safer and easier.

Arthur Kornberg
Stanford, California

Preface

Safety was the paramount objective when phosphate fibers were invented. Safety for workers and consumers is the only technical reason anyone should consider launching a program to produce a substitute for the serpentine minerals usually referred to as "asbestos." There is no reason to expect that synthetic fibers can outperform natural mineral fibers nor is it expected that they can be manufactured more economically. Modern societies require mineral fibers if we are to continue to advance in areas such as human safety, nutrition, health, comfort, transportation, and longevity.

It has been demonstrated through centuries of human exposure and intimacy with inorganic polyphosphates that they are jewels of safety in an environment where most common elements of the periodic table are viewed as potential threats. The underlying reasoning that made phosphate fibers a reality came from a belief that "asbestos-like" crystals could be grown from polyphosphates. If phosphorus atoms could be substituted for silicon atoms in molecules similar to serpentine minerals, a safe mineral fiber should be possible. Polyphosphates are indigenous to life from conception to death and their role in life is well established. Silicate minerals, though usually benign, are not life giving.

Challenges to produce safe mineral fibers were undertaken by a group of dedicated, highly qualified scientists, engineers, and technicians. Results were more successful and satisfying than any of us dared hope for when the crusade was launched in an aura of noble intentions and promises of a safer world for everyone. The only negative aspect of this project resulted from activists claiming to champion all things wholesome while attempting to disembowel all efforts to bring forth newer and safer substances for the benefit of everyone. Phosphate fibers were researched and developed under constant harassments of shameless ignorance, questioned intentions, and threats of lawsuits.

Much phosphate fiber research and development continues in many parts of the world today. Their properties are remarkable and many uses have been discovered that have nothing in common with "asbestos." Bone implants, food uses, hard surface cleaners, and beverage filters are but a few noteworthy examples.

The Phosphate Fiber Honor Roll

The following Monsanto employees made significant contributions to the Phosphate Fiber Project. Their contributions are recognized and highly appreciated. We express our thanks as well to the dozens of others who contributed and supported the Phosphate Fiber Project.

Adamove, J. E.
Apostolou, D.
Athmer, C. J.
Ballard, V. L.
Beckman, M. R.
Bolte, H. F.
Brooks, W. W.
Brostowicz R.
Cannon, J. A.
Cohen, L. A.
Combs, J. F.
Crutchfield, M. M.
Domahidy, S. E.
Downes, J. E.
Eaton, D. R.
Edwards, D. A.
Eggert, C. R.
Fallert, D. L.
Flowers, L. J.
French, J. C.
Gard, J. K.
Gledhill, W. E.
Golland, D. I.
Gossage, T. L.
Griesbauer, R. C.
Hall, L. J.
Hansen, R. L.
Haynes, R. L.
Henn, A. R.
Hezel, S. A.
Hill, J. C.
Hines, D. B.
Hinkebein, J. A.
Hogue, R. D.
Jany, J. R.
Johannsen, F. R.
Kaelble, E. F.
Kidwell, R. L.
Kilbourne, H. W.
Knox, W.
Krick, G. O.
Lannert, K. P.
Lee, P. E.
Li, A. P.
Mahoney, R. J.
Manno, A. L.
Mappes, G. W.
Matzner, E. A.
McDaniel, W. C.
McEwan, G. J.
Monroe, L. D.
Monzyk, B. F.
Mosher, R. G.
Myers, C. A.
Nair, R.
Narciso, J. J.
Newton, P. E.
Ngo, T. M.
Niewoehner, L. E.
Novell, B. A.
Oostendorp, D.
Payne, J. H.
Peters, J. A.
Puronen, E. O.
Rawlings, G. D.
Rinehart, W. E.
Roper, C. B.
Rowell, R. R.
Sakata, M.
Schifferdecker, R. L.
Schrichte, W. C.
Seder, A. W.
Shearing, F. D.
Shiotsuka, R. N.
Short, R. D.
Silverman, B.
Stahlheber, N. E.
Tierney, P. A.
Trehy, M. L.
Wang, J. T.
Wearing, J. R.
Wilson, A. G. E.
Winkel, M. W.
Wolfe, T. C.
Wulfert, K. J.
Zimmerman, V. C.

Edward J. Griffith

Monsanto
St. Louis, Missouri

Acknowledgment

The author wishes to express his sincere appreciation to the following experts, who contributed their knowledge:

Howard J. Buttery, Ph.D.
Technical Manager, 3M Encapsulation Products, 3M Center
St. Paul, Minnesota

Author Kornberg, Nobel Laureate
Department of Biochemistry, School of Medicine, Stanford University
Stanford, California

Raymond C. Loyer, Attorney
Law Department, Monsanto Company
St. Louis, Missouri

Contents

1. Fibers, Phosphates, and Public Protection

2. The Limiting Element for Life on Earth

4. Ropes, Strings, Threads, Fibers, and Molecules

5. Melts and Phases

6. Laboratory Experiments

8. The Many Uses of Phosphates and Fibers

9. Product Safety in the Climate that Phosphate Fibers Were Introduced: How Much Testing Is Enough?

10. The Future of Phosphate Fibers

1

Fibers, Phosphates, and Public Protection

1.1. Editorial Comment

This book tells, in the simplest terms the author is capable of writing, the day-to-day experiences of a group of dedicated scientists. Their mission was to bring to the world a safe, inexpensive, mineral fiber that everyone could use without fear. It also teaches the background science and engineering required to bring this project to a technical success, only to have the project abandoned as a direct result of its initial reason for existing. Societies worldwide have become so afraid of "asbestos" that any mineral replacement for "asbestos" is viewed as potentially harmful, and is usually attacked immediately after it is announced.

Too often a product is found guilty before it has had a fair trial. Unfortunately, the laws of logic make it impossible to prove that a product is innocent, once accused. Those claiming to be injured by "asbestos" can obtain large sums of money as compensation for almost any illness, real or imagined. This is not to imply no one has been injured, but merely to say it is almost impossible to protect one's self from claims of injury.

Throughout this work there were several fundamental focal points that are believed to be scientifically and technically sound. In dealing with the safety of any product, nothing is absolute and neither are these focal points. The interaction of a product and a host is so complex that no one can predict what will happen under all conditions to all hosts. As an example, most diners will have no adverse reaction to shellfish while others will become seriously ill when eating only small quantities of a substance to which they are extremely allergic. One shrimp was sufficient to send a friend to the hospital each time she has ever attempted to eat one. When a diner becomes ill after eating sanitary shellfish, is this a test of the shellfish or the diner? No product is completely safe for everyone, under all conditions, and no tests or sets of rules can anticipate all circumstances. Nevertheless, some guiding principles are a requirement.

The following guiding principles were used as a checklist. It is intended that they will add some continuity to the stand-alone chapters.

FOCAL POINTS

1. Fibers enter a human host by inhalation or ingestion. All other routes can be neglected in this work, except toxicity tests where fibers may be inserted into any orifice or incision.
2. Inhalation is the primary means of concern of this work. Protection of lungs is the ultimate aim. Toxicity resulting from chemically benign fibers is confined primarily in the respiratory tract unless metastasis has occurred within the host.
3. In healthy hosts, inert fibers that are too large to enter alveoli are judged to create no significant health hazards if inhaled in small quantities.
4. Unless chemically toxic, fibers that are known to be harmful to resident hosts may not be immediately toxic to these hosts. Months or years may pass before toxic symptoms appear in these hosts unless the exposure is intense.
5. Fibers that become attached to membranes are more likely to cause problems than those that do not. These can become points of long-term irritations.
6. Inert fibers that are quickly flushed from a host are not usually considered to cause health hazards.
7. Fibers that are composed of components that are residents in a host organism will be less inclined to be threatening to a host. These substances are less likely to be treated as foreign substances by a host. Polyphosphates are an example of such substances. Adenosine triphosphate is a notable example.
8. Fibers that are degraded or transformed to nonfibrous solids or nontoxic solutions are not considered toxic to a host.
9. Refractory fibers are more inclined to cause problems in human hosts than are fibers that can dissolve or degrade. Polyphosphates are the ultimate in biodegradable substances. They are unique among mineral substances.
10. Fibers must be strong and sufficiently resistant to attack to be useful in applications where mineral fibers are traditionally used.
11. Crystalline fibers are usually more stable than their amorphous counterpoint fibers and the properties of crystalline fibers are more predictable.
12. Polyphosphates are thermodynamically unstable in the presence of water and will degrade slowly owing to hydrolytic degradation.
13. Polyphosphates biodegrade more rapidly by hydrolytic degradation in the presence of enzymes created in living systems.

14. Phosphates are nutrients in all living systems and life cannot flourish or propagate without phosphates. More than 25% of the weight of DNA is orthophosphate.
15. Polyphosphates cannot be burned and do not support combustion.
16. Sodium and calcium polyphosphates have been used in foods and potable water supplies for many years and are not toxic when inhaled or ingested.

These focal points will be referred to from time to time throughout this work. They are the foundation on which the cornerstones of the Phosphate Fibers Project were laid.

"Asbestos" is no longer front-page news. Writers have exhausted all factual issues. Twenty years ago the idea was promoted that "asbestos" was a major threat. It was implied that everyone should expect to die eventually as a result of lung cancer unless "asbestos" is banished from the face of the Earth. The reporting was frightening and convincing. I believed that workers who handled mineral fibers were in danger of being uselessly sacrificed. This belief caused me to "bootleg" research on a new, safe alternative to "asbestos," Phosphate Fibers.

To be a scientist or a manager in a large corporation requires planning. In most large companies, very little happens during a year that was not anticipated during goal setting and planning at the start of a year. This is necessary, but when goals have been approved from top management down to laboratory scientists, established plans are difficult if not impossible to change. When promising new targets of opportunity suddenly arise that were not anticipated when goals were approved, they are usually explored without direct corporate support if anything is ever to happen. This is an accepted part of an unwritten plan. Such was the case for the Phosphate Fiber Project. Time and money were spent that had not been approved before the project was introduced to management. All scientists are aware that they are free to make short-term probes on their own authority.

During the last half of the 1970s I was visiting a friend, Sidney Campbell, in New Orleans, Louisiana. He was a distributor for the Johns Manville Company. Sidney's "asbestos" plant had just been condemned and closed by order of the Environmental Protection Agency. "My plant has never caused any health problems. All health officials admit there are no health problems with my workers. My family and I worked with "asbestos" all our lives. My father and I are roofers. I have handled "asbestos" all my life. Now I must fire many of my workers because they might get sick." Chrysotile was the fiber used in the roofing and it will be seen that it must be compared with other mineral fibers to determine how much hazard is associated with chrysotile. All "asbestos" was not created equally and there were cement pipe manufacturers in New Orleans who may have caused problems.

Until that night, I had never given much time or thought to "asbestos." I knew very little about "asbestos." I could think of no person I knew who had been harmed by "asbestos." Maybe there was one, but I was not certain he had ever handled any "asbestos." If it is widely believed that "asbestos" is a killer, this is almost as bad as though it were. "Something needs to be done," I thought.

Next day I drove almost 800 miles back to my home in St. Louis, Missouri. Interstates were still relatively free of traffic. This allowed me long hours for thinking. I am the kind of environmentalist who wants to solve problems rather than ban everything that is effective. I am thinking, "Humanity needs mineral fibers. Can I do anything to help solve the problem? No one can question the safety of polyphosphates. Can I make a new kind of "asbestos" where the silicon atoms in the fibers are replaced by phosphorus atoms? It is surely worth a try! I am resolute. It is a worthy goal. I will try."

* * *

In recent years many well-meaning persons have warned all of us in the public at large of every conceivable hazard humanity has ever confronted and many that do not exist, except as models in imperfect computer programs and strange toxicological protocols. It is true, almost everything on Earth is hazardous in one way or another. Many of my older friends have become too frightened to enjoy living in the most wonderful era humanity has ever been privileged to experience. They have read or heard from an authoritative reporter that everything is potentially harmful. It probably is to some degree or other.

Powerful monarchs of the past could not command for any price the slaves we instantly summon and command at the mere flip of a switch. We can invite the world's greatest entertainers into our living rooms. The best musicians are welcomed in living stereo. If out of town, we fly home on a magic carpet of splendor; Ali Baba could not dare dream of this wonder. Shiny horses meet us at the airport and deliver us to our front doors. Our house slaves control temperatures winter and summer to within 1 °F by a preprogrammed thermostat that will answer to a telephone call. It is not merely the rich and famous, but most of us who have these fantastic gifts. Caesar, Lincoln, Napoleon, or Washington would have gladly traded places with even the less fortunate of us. Poverty-level citizens in America control more power than monarchs of only a few years ago. Almost everyone has telephones, automobiles, central heat and air conditioners with thermostats, electric lights, cooking stoves, and some form of medical care, to mention but a few.

Sadly, too many of our friends and neighbors have become frightened of everything from their food, to their air, to their drinking water supply. All supermarkets are overstocked with bottled water that does not contain the proved benefits of fluorides, because we have frightened our people of their basics of life. I am unable to convince some friends of the dangers of continuing to drink distilled water as their only drinking water supply. Distilled water is very corrosive to most metals and even glass. It has been recognized for many years that people who live in areas where their water is very soft are inclined to have kidney stones.

Most of the childhood and young adult killers of the past, such as poliomyelitis, malaria, osteomyelitis, diphtheria, typhoid, appendicitis, and many more, have been pushed out of our minds as though they never existed. Warnings have been so overstated by those wishing to protect us, parental cautions become callous threats. Each protector endeavors to see that we do not ignore his important, unique, warning consequently leading to overstatements.

One purpose of this book is to share with readers how a research project came to life to end a problem I believed to be a serious threat to everyone. I accepted the warnings as stated. As I learned more and more about the hazard, I became less and less convinced that the "truth" I had been led to believe was real. There is a hazard. The hazard is similar to the one we confront when we cross busy intersections. There is a hazard, but few if any books are written entitled *The Trauma of Crossing Busy Intersections.*

1.2. Birth of a Research Project

In New Orleans I had been working on an ammonium nitrate explosive. A few years before, I had invented a stabilized ammonium nitrate that was magnificent for use in farming, but would not absorb enough oil to make it a good explosive. I had been requested by one of the other divisions of our company to help find ways to cause agricultural grade, dense ammonium nitrate to function as an explosive. Most industrial blasting, for mining, highway construction, and the like, is done with ammonium nitrate, rather than dynamite. Ammonium nitrate is safer and less expensive. A truly satisfactory solution to the oiling problem was never found, although it occurred to me while writing this paragraph for you how it could be done. That is another story.

When I returned to the laboratory in St. Louis the nitrate paper work had to be handled first. A few days later, when I found some spare time, it did not take long to put two hundred grams of phosphate in a platinum dish and heat it to 800 °C in our laboratory furnace. Did this initial idea have any merit? Not much, but this was to be expected. Very few raw ideas work well when first tried. Each day I would try again and everyday some small improvements were seen. (Later you shall learn how Professor Abe, of the University of Nagoya, Japan, took this initial idea and some brilliant research to produce a product that should be of great value to everyone in the future.)

After about one month of semifailures, a sample was prepared that had some properties I had hoped to find. The sample was similar to "asbestos" in appearance and feel. I called my friend, Dr. Marvin Crutchfield, a world renowned detergent scientist, from the lab next to mine. I solicited his impression of the new fibers. Marv smiled, "By George, I think you have done it."

It was time for me to approach my management. I had completed the work on ammonium nitrate for the other division and timing was right. I told myself, "It is time for salesmanship, not science. How can I get my boss to recognize that this is a project that he should support?"

I called my manager and asked if I could meet with him for a few minutes. He said that he was busy at the moment, but I should come to his office about one hour later. I walked into his office carrying two petri dishes with white fibers piled high in them. The fibers were white, soft, and fluffy. I told my boss of the need for a safe "asbestos" and that I was convinced that I could prepare a safe substitute for "asbestos" from polyphosphates.

As I talked I kept fingering the white fibers in the dishes, but I did not offer them to him. After about twenty minutes it was obvious that he was becoming more and more interested in getting his hands on the fibers. It was time for me to make my play for support. Not even the best of scientists can work alone. Everyone needs help and everyone must be a part of the team. I pushed the two dishes across his desk for him to examine hoping he would be impressed. I needed him before I could move further on the project.

Dr. John Payne fingered the samples. He rolled them between his fingers. He allowed them to fall back into the dish after he examined small pinches of the pretty materials. Finally he asked, "Which one is the "asbestos?"

Triumphantly I answered, "Neither. They are both polyphosphates!"

He was hooked. He recognized that he was incapable of knowing a polyphosphate from "asbestos" when the polyphosphate was properly prepared. The Director responded to Dr. Payne in much the way Dr. Payne had responded to me. There was no way to stop the Phosphate Fibers Project. We were on our way even if it was not the start of a planning year. If a project looks good enough, simple rules are made to be broken.

* * *

One of the more fascinating challenges to confront an experimental scientist is to attempt to recreate and improve a superb natural product. Conversion of lead to gold and synthesis of diamond from graphite have intrigued experimenters for centuries. Both conversions can be accomplished today, but neither has fulfilled dreams of converting ordinary plentiful substances into rare objects of great economic value. There is small doubt that if these conversions could be made cheaply, great profits for a small quantity might soon vanish. But, the outstanding properties of these materials will still have great economic value as very useful commodity items. This will surely be of greater cumulative value to humanity when it is accomplished.

Serpentine minerals have rivaled all other natural products for unique properties, exquisite detailed structural beauty, and service to humanity. It is reported that one night more than one thousand years ago, Charlemagne, Emperor of the Holy Roman Empire, triumphantly snatched a tablecloth from his banquet table, regally hurled it into a raging, crackling, open fire with no damage to his tablecloth. Since that night these mineral fibers have reigned supreme. To astonish his guests Charlemagne demonstrated that even his tablecloth did not dare burn without his permission. Whether or not this story is true is questionable because most of the records have been destroyed. Also, the act is attributed to several different men. There is no doubt that the Romans did wrap their dead in robes of "asbestos" and even Alexander The Great may well have been the first to demonstrate his powers with a tablecloth.[1]

Minerals which can be separated into snake-like fibers are named by mineralogists as serpentine. Serpentine minerals have been of unquestioned value while being exposed to humanity for thousands of years. Practically all humanity, in advanced nations, has been immersed in these fibers for decades with neither protection nor thoughts of any hazard. Only within recent years has their service been questioned. If serpentine minerals even approached being as hazardous as they are claimed to be, every person on this planet should have died of lung cancer, mesothelioma, decades ago.

There is adequate self-evident history to question the cause for a near-hysteria surrounding fibers in recent years. Buildings have been gutted, mothers have refused to send their children to schools, public buildings have been abandoned, church members have refused to attend church,[2] all because "asbestos" was believed present. On July 20, 1986 E. De Wolfe wrote an article for the Los Angeles Times, *Asbestos in the Home: Avoid Risk of Doing Too Much,* in which he notes that often it is wiser just to leave the mineral undisturbed.

The Golden Gate highway bridge was closed, creating a monstrous traffic jam, merely because some building materials, believed to contain "asbestos," fell on the bridge! One minister reported to me that members of his church had threatened to sue him, because they believed that there was "asbestos" in ceiling tiles of his church. The Rockwood School District of Missouri carpeted all floors because there was "asbestos" in the floor tiles. Floor tiles have never been considered a hazard.

No other fibrous substances approach the wondrous physical and chemical properties of these *very natural* serpentine minerals. In retrospect, had I known more about serpentine minerals when I initiated the Phosphate Fibers Project, it is doubtful whether I would have had enough courage to attempt to prepare a phosphate replacement analogue of these minerals. The more I learned of these fibers, the more wondrous their properties became.

Unfortunately, there are those who relate natural to safe, synthetic to harmful. "Asbestos" is as natural as substances exist on Earth, although it does seem to possess supernatural properties. Humanity had no part in bringing "asbestos" to Earth, nor is humanity capable of removing it from Earth or destroying most of that which is here. Many educated persons have been misguided to believe "asbestos" is a synthetic substance. They believe "asbestos" is an undesirable substance, for no better reason than they mistakenly believe it is "unnatural." "Asbestos" must be banished from planet Earth.

"If seven maids with seven mops
Swept it for half a year,
Do you suppose," the Walrus said,
'That they could get it clear?'
'I doubt it,' said the Carpenter,
And shed a bitter tear."

Lewis Carroll
Through the Looking Glass

It is unfortunate that a variety of useful serpentine minerals was lumped under a generic name "ASBESTOS" and treated as though "asbestos" is a single substance. In a relatively short period, "asbestos" was labeled an evil mineral. In my judgement, some forms of serpentine mineral fibers are indeed harmful when improperly handled, but other forms of "asbestos" are as safe as most particulate substances encountered on Earth. To breathe large quantities of dust from any source is surely damaging. There is adequate proof coal dust produces "black lung" and emphysema, textile fibers cause "brown lung," mill fever, (byssisois) and emphysema, and some forms of serpentine mineral dust cause asbestoses, emphysema, and mesothelioma. These are well documented facts.[3]

Highly developed societies have grown so dependent upon fibers that it is almost impossible for these societies to exist and progress without fibers. Health concerns have caused some nations to abandon even minimal risk of mineral fibers. These nations are depriving themselves and their citizens of products derived from these minerals' unique properties. There are challenges with many useful modern substances. We must learn to use these substances wisely rather than attempting to banish them from Earth. Often there is no substitute that can perform reliably. Too often when substitutes replace a product with a long established history, little more is accomplished than to trade known troubles for unknown troubles. When functional substitutes for existing functional products have been used for a sufficient time, they too will have properties that are undesirable in ways yet to be discovered. The mere fact that they are useful and used in large quantities will cause them to be judged good by some and bad by others.

It is not the purpose of this book to enter a never-ending debate whether or not "asbestos" is harmful or safe. After thousands of hours of court hearings, lawsuits, laboratory studies, epidemiological surveys, clinical studies, and billions of dollars for all manner of cost, safety issues remain cloudy. There is no doubt that some persons have been harmed, but whether or not "asbestos" constitutes a hazard to our general public on a casual contact basis is indeed questionable. Even a well-meaning medical profession's judgement is probably distorted; they encounter those who are already ill. A positive result for someone testing a substance for harmful properties is to identify the harmful properties. Few addresses are presented to scientific meetings of great societies of "pith and moment" on the harmlessness of "stuff."

1.3. Phosphate Fibers

From the moment of conception of phosphate fibers, during all research and development of phosphate fibers, it was our expressed purpose to deliver a safe, useful, inexpensive mineral fiber to society. It cannot be questioned that our mission was noble. A profit motive existed; a corporation ceases to exist when there is no profit, but profit motive was not the driving force behind this project. Because of our mission it would be both deceitful and cowardly to ignore safety issues that deal with natural "asbestos." These issues will be considered for comparison to phosphate fibers when it is deemed necessary. If there were no safety issues a Phosphate Fibers Project should never have existed. Properties built into phosphate fibers, to insure their utility and safety, were chosen to overcome the problems associated with "asbestos." The properties of the phosphate fibers must be compared with natural serpentine minerals, both negatively and positively, if this book is to serve a useful purpose.

From time to time I will inject some of my views concerning usefulness and safety of mineral fibers, recognizing that it is impossible to become expert on all issues surrounding mineral fibers. After two decades of intense interest in this subject it is difficult to avoid biases, many of which are based upon impressions gained by studying only a part of all information that has been published. To be conversant with all published literature dealing with this subject should be a full-time occupation. In my case, the "asbestos" issues are definitely sidelines to my primary occupation.

Few persons on Earth should have more reason to damn mineral fibers and their industry than I. Hopefully, this presentation will treat both sides of this question, with objectivity and fairness. After dealing almost daily with an ever increasing number of rules, regulations, and roadblocks that stand in my way to deliver new substances that I believe could be a service to everyone, I am frustrated on a number of issues. I find that I first vent my frustrations by writing them to get them out of the way, and then I delete them to spare you my concerns. My apologies, in the event I do an inadequate job of removing my disdain at times.

One purpose of this book is to teach of efforts and new knowledge learned by a group of talented, highly dedicated men and women, determined to bring to all societies a safe mineral fiber substitute. Few projects in any large corporation have ever received more instantaneous support from the company's scientific community. My first short introduction of the concept to our technical community received a standing ovation. Members of the Phosphate Fibers Group wore their responsibility with open pride. The Fibers Project was believed by those contributing to the project to be worthy of their finest efforts. They spontaneously gave their best to overcome difficult problem after difficult problem, any time the project seemed in danger of being terminated. Some critics in early days expressed their sincere belief that the project was "impossible." One engineer labeled the process, "Griffith's Hell's Fire Process." Temperatures needed to make fibers were very high and containment of phosphate melts was a major challenge in all early designs.

The Phosphate Fibers Group approached their work with an aura of a noble crusade. Many more scientists and engineers requested to join the project than could be accommodated. Few, if any, employees in a company of more than 50,000 at the time were unaware of the Phosphate Fibers Project and followed it with almost daily interest. They believed that when the project ultimately became successful, a new fiber would bring peace of mind to mothers, who of necessity had to leave their children in daycare centers or schools, believing they were exposed to unnecessary hazards. A second paramount purpose was to bring assurance of a safe workplace for workers, who daily exposed themselves to fibers, to supply superior goods and services for all the rest of us. There was no doubt that pride came from a steadfast belief, "We are engaged in research and development of extreme importance for our families and future humanity!" It is true!

If you are confused about my feelings toward "asbestos" safety, you have every right to be. Today, I meet this subject with mixed emotions. There seems to me to be no doubt that we overreacted to potential threats "asbestos" poses for humanity. Conversely, I have no doubt that a mineral fiber whose safety is accepted by all responsible persons is still needed. But, you will learn that I believe that nothing on this planet is completely safe and that it would be impossible to prove that it is safe even if such a substance could be known to exist.

1.4. What Is a Fiber?

At first thought, "What is a fiber?" may seem to be a nonquestion. Everyone knows what a fiber is. A critical definition of a fiber is not easily formulated. Most dictionary definitions are ambiguous. For example, here is a dictionary definition: "Fiber or fibre: a thread or a structure or object resembling a thread: as a (1) : a slender root (as of a grass) (2) : an elongated tapering thick-walled plant cell void of maturity that imparts elasticity, flexibility, and tensile strength b (1) : the axis cylinder of a nerve cell with its sheath (2) : one of the filaments composing most of the intercellular matrix of connective tissue (3) : one of the elongated contractile cells of muscle tissue c : a slender or greatly elongated natural or synthetic filament (as wool, cotton, asbestos, gold, glass or rayon) typically capable of being spun into yarn 2 : material made of fibers; specif. : VULCANIZED FIBER 3 a : an element that gives texture or substance b: basic toughness: STRENGTH c : essential structure"—*Webster's Seventh New Collegiate Dictionary*, G. & C. Merriam Company, Springfield, Massachusetts, 1969.

The above definition is more examples of fibers than a definition of a fiber. It supports the fact that not only is "asbestos" difficult to define, but fibers themselves are also difficult to define, for other reasons. Hopefully, a better definition can be based upon aspect ratio. Aspect ratio is defined as a ratio of the length of an object to its diameter. A fiber can be defined arbitrarily as an object with an aspect ratio of 10 or greater. (Some choose to define a fiber as any object with an aspect ratio of three or greater.[4] By this definition most crystalline substances are fibers.) One deficiency of a definition based upon an aspect ratio of 10 or more is that no maximum diameter is specified. Therefore, a fence post, a utility pole, a railroad rail, a longboat, are all fibers by an aspect ratio definition. This does not serve our needs for a definition. Few readers will agree that a railroad rail is a fiber or that an elephant's rib is a phosphate fiber. In fact, an elephant's rib is a calcium phosphate fiber, albeit a large one.

To most readers the word fiber probably brings to mind a thread-like substance that is highly flexible and is easily twisted or bent. Since many substances classed as mineral "fibers" are not thread-like, but are acicular, rigid needle-like bodies, an ability to be, or not to be, easily flexed probably should *not* be included in a definition. However, if rigidity is accepted as a characteristic of some types of fibers, then sewing needles, nails, and straight pins fall under one definition of a fiber, because most exhibit aspect ratios greater than ten and they are rigid.

In this work a fiber will be loosely defined as a thread-like body with an aspect ratio greater than ten, but having a maximum diameter no greater than 0.1 cm. This is arbitrary, but similarly there is no definitive definition that differentiates a thread from a rope. Is a thread a small rope? Or is a rope a large thread? In this work only fibers possessing diameters within the definition will be considered. Few, if any, important, fibrous substances will be excluded by this definition. If there is a problem, it will be that too many substances will be classified as fibers when they probably should be defined in some other way. A sewing needle will be defined as a fiber if it has a diameter less than one millimeter, and most are less than one millimeter in diameter.

Fibers can be divided into two general classes: crystalline or amorphous. These may be further divided into four more classes: natural organic, synthetic organic, natural mineral, and synthetic "mineral" fibers. Perhaps the word mineral should be reserved for natural substances only, but definitions here have also become so loose that all liquid and solid inorganic substances are referred to as minerals. In usage adopted for this book an effort will be made to differentiate between natural and synthetic fibers.

Most organic fibers are amorphous or semicrystalline. Mineral fibers are highly crystalline or amorphous. Most natural mineral fibers are highly crystalline while most synthetic mineral fibers are amorphous (fiber glass, spun slag, or mineral wool). Carbon fibers are a notable exception. They can be obtained by decomposing organic fibers, such as those made by charring polyacrylonitrile. Carbon fibers can be crystalline.

1.5. Nature's Fibers

Fibrous materials have been indispensable to humanity from the earliest tribes to modern global societies. National economies have depended upon fibers such as cotton, wool, silk, flax, hemp, and wood as their primary items of commerce. Most natural fibers are intentionally grown in a selective agricultural project and are harvested in much the same way other farm products are cultivated and harvested.

Natural organic fibers have several properties that can limit their usefulness in applications where mineral fibers are often used. Since natural organic fibers were derived from some life process, their association with water is to be expected. They are predisposed to either absorb water from their surroundings or to lose water to their environment, depending on conditions. Imbibing or losing water by a fiber can change its physical dimensions, causing it to either swell or shrink. Often, a change of physical properties can detract from a fiber's usefulness and the attractiveness of a finished product made from these fibers. Conversely, many inexpensive humidity gauges depend upon shrinkage and expansion of horse hair in response to changes in vapor tension of water in air. These gauges give reasonable estimates of relative humidity as temperatures and vapor tensions of air change.

Most, if not all, natural organic fibers are subject to some form of biological attack. This can be both positive and negative, depending upon an intended use of a fiber. Natural organic fabrics are much less likely to collect in our environment than some of their synthetic counterparts. Although our environment might be improved by returning to natural fibers in some of their former applications, fewer people would be alive to enjoy their environment. Until World War II American tire manufactures were required, by government edict, in a free enterprise system, to use only cotton cord in tires. This was caused by what was known, at the time, as the "Cotton Block" in the United States Congress. Southern congressmen imposed cotton on the peoples of the United States to protect their cotton picking constituents in the Cotton Belt. There was a similar, almost disastrous, "Rubber Block" requiring that only natural rubber could be used to manufacture tires. It is not clear whose farmers this law protected.

Blocks were supposedly done to help cotton farmers, though simultaneously, limits were placed upon quantities of cotton that could be grown. Cotton cord tires were a menace to all society. They were biodegradable and often failed catastrophically at moments of disaster. Few reasonable people would desire that we return to days of "natural cotton" tire cord or "natural rubber" automobile tires. These tire's life expectancy was about twenty thousand miles, under the best of conditions, but they were highly susceptible to flats and blowouts, shortening their useful lives even more. It is fortunate that some oil companies had the foresight to develop synthetic rubber from oil, even if Congress did not recognize how dependent the United States was on a product we were incapable of growing in the United States.

The range of temperature stability of natural organic fibers limits their utility and most are highly flammable, having caused death to thousands of their users. Although statistics are unlikely to be very reliable, more deaths and disabilities have occurred because natural organic fibers are flammable than have ever occurred from "asbestosis" and lung cancer (mesothelioma). In 1967, in the United States, about 1,700 persons died from bedding fires alone.[5] It has been estimated that approximately 5,000 persons in the United States die as a result of fires.[6] A similar number were expected to die from all forms of mineral fiber contacts.[7] Every year more than 2,000,000 Americans die from one cause or the other. Too often, data of exact types required to estimate relative hazards are not collected and saved. This can lead to misplaced attention being directed to a lesser hazard while ignoring impending disasters.

Synthetic organic fibers have been used to replace mineral fibers in some application since mineral fibers have come under attack. Natural serpentine fibers are mined from Earth and are therefore relatively inexpensive when compared to synthetic organic fibers. Aramid polymers, Kevlar®, have been substituted for serpentine minerals in automobile brakes and seem to be reasonably effective, but expensive.

1.6. Mineral Fibers

Throughout this book I shall refer to serpentine minerals by their common mineral names. The word "asbestos" will be reserved to emphasize the word's misuse. Four mineral fibers will dominate our attention. These minerals are:

1. Chrysotile. This is a magnesium silicate hydrate very similar to talc, chemically.
2. Crocidolite. This is an amphibole crystal that is similar to chrysotile in bulk appearance.
3. Amosite. This is an acicular (tiny needles) amphibole and has a very bad reputation as a health hazard. It is probably well deserved.[8]
4. Tremolite. This is an amphibole crystal with properties similar to both crocidolite and amosite.

There are many other fibrous minerals that could be listed ranging from the gem stone, tourmaline, to wollastonite, or a clay, attapulgite. We shall discuss these fibers in more detail in Chapter 4.

One early bias confronting this author when the Phosphate Fibers Project was initiated, but had not yet been sanctioned by corporate management, was fiber lengths. It was pleasing to picture one's self parading down laboratory halls, holding at one end a bundle of long, silky, shining fibers, while trailing four feet behind, like angel's hair of a comet's tail, streamed our new pride. It came as something of a disappointment when I learned that almost no one wanted long fibers, for anything. Most customers wanted a mineral fiber about 0.5μ in diameter and maybe 100μ in length. This would make our task much simpler, but not as much fun.

Many properties of fibers must be considered when choosing a fiber for a particular job. Often specifications required by a customer are lengthy and demanding. Fibers are used to insulate items as small as appliances and as large as office buildings. In other applications a fiber's insulating properties may have no value, while strength of a fiber is of primary importance. Surface area can become crucial when filtering resources are investigated. It is obvious that those who design and formulate products require as long a list of available properties as it is practical to maintain. Few, if any, substances are either all good or all bad, particularly if a substance is functional. It is usual that some bad properties must be minimized to obtain benefits of other desirable properties needed in a particular application.

Most mineral fibers are silicates and are thermodynamically stable under all but extreme conditions. They are immune to attack by biological agents. They can often withstand wide variations in temperature, pH, and other stresses without decomposing, having been exposed to a wide variety of conditions for millions of years. This resistance to degradation and change, under most atmospheric conditions, is responsible for both their usefulness and abandonment.

1.6.1. Amphiboles and Chrysotile

Two general types of silicate minerals are usually classified under a poorly defined misnomer, "asbestos." Scores of fibrous minerals can be lumped under this term when both chemistry and mineralogy are employed for identification, particularly of amphibole minerals. Amphibole minerals are the most chemically diversified. They can be differentiated from other silicates by their molecular structure. Amphibole crystal's molecules are oxygen cross-linked double chains.

```
    O⁻  O⁻  O⁻  O⁻  O⁻  O⁻  O⁻  O⁻  O⁻
----SiO SiO SiO SiO SiO SiO SiO SiO SiO-
    O   O   O   O   O   O   O   O   O
----SiO SiO SiO SiO SiO SiO SiO SiO SiO-
    O₋  O₋  O₋  O₋  O₋  O₋  O₋  O₋  O₋
```

A second class of silicate mineral called "asbestos" is a serpentine mineral, *chrysotile*. It is a magnesium silicate hydrate, $Mg_6[Si_4O_{11}(OH)_6]H_2O$,[9] with a chemical composition similar to talc, $Mg_3Si_4O_{10}(OH)_2$.[10] There is at least a plausible argument that *chrysotile* should not have been included with other minerals named "asbestos" from chemical, health, and legal considerations.[11] There is small doubt that the amphiboles are much more likely to cause difficulties. A study of two cement pipe plants in New Orleans revealed some interesting results. In one plant only chrysotile was used. In a second plant both chrysotile and amosite were used. It was found that there were four times as many lung cancers in the plants that used both chrysotile and amosite as in the plant that used only chrysotile.

More than ninety percent of "asbestos" used in the United States was *chrysotile*. It is probably the least hazardous of all substances called "asbestos" and may be only slightly, if any more, hazardous than its sister, talc.

1.6.2. Talc

It is difficult to understand our reaction to talc considering the public's concern for "asbestos." Talc has been dusted on surgeon's gloves, powdered on baby's rears, perfumed for after baths, and dashed on our faces for most of a lifetime. Perhaps it is because we are so intimate with talc that we treat it as we did saccharin. All of us were so familiar with saccharin we decided we would use it regardless of who called it a carcinogen. We shall return to this point many times as we attempt to understand why we become so determined to rid ourselves of one substance while a second substance that is even worse is ignored.

Talc is in the same acceptance class as saccharin. It is another case of familiarity breeding acceptance. "Talcum has never caused me a problem, why should it cause me a problem now?" It has been claimed for many years that some talcum powders contain several percent *amphibole* "asbestos."[12] This is strange because, as noted above, the chemical formulas of talc and chrysotile are very similar. Neither talc nor chrysotile are amphiboles. It would be expected that if talc were to contain a fibrous component it would be one of similar composition, but this is not the case. The structure of chrysotile is almost unbelievable and perhaps talc does not provide a proper environment for this crystal growth.

James E. Lockley, M.D. wrote one of the more informative articles ever written about mineral fibers. He traces the medical history of talc back to 1896 when it was first reported to give respiratory problems similar to the ones attributed to amphiboles.[12] Talc is believed to cause ovarian cancers in some women who have used it on sanitary napkins.[13] Why does no one seem to care? Should we care? I don't know, but I am always concerned about our complete lack of consistency in the way we present information to the public. "Wolf!" has been cried so many times that the public at large is beginning to believe nothing, but to fear almost everything.

The above coverage of talc is not intended to raise new suspicions about talc. There are too many substances to alarm the public's concern without adding another. Additionally, the talc industry has done much to mine talc that is not contaminated with mineral fibers, particularly for talc used for cosmetic reasons. Mineral fibers can no longer be observed in most of the products being sold today.

1.7. Amorphous Phosphate Fibers

Phase transitions will be dealt with in more detail later, but when a glass crystallizes it undergoes a phase transition from an amorphous state to a crystalline state. When water evaporates it undergoes a phase transition from a liquid state to a vapor state, or when it freezes it undergoes a transition from a liquid state to a solid state. When a product is manufactured it usually has desired specified properties. If this product suffers a phase transition its properties will change, often suddenly and catastrophically. One does not wish to ride on an airliner in which the metals of the wings are undergoing a phase transition. This is called metal fatigue and in some cases the wings are likely to go their separate ways.

Amorphous spun phosphate fibers can create effectiveness problems when used for reinforcement if they are inclined to devitrify. If crystallization is catastrophic, an entire fiber can be converted to a string of fine powdered crystals as a phase transition occurs. If decomposed fibers are exposed to an environment, then a dusting problem may also result. Amorphous fibers are also much more inclined to be hygroscopic than are their crystalline counterparts. Usually, a crystalline fiber is superior to an amorphous fiber, when short reinforcing fibers are compared for utility and a reliable long life.

Numerous long-term projects have been initiated in an attempt to spin amorphous phosphate fibers. Until now they have been unsuccessful for long-term applications, but can be safe for the same reasons that crystalline phosphate fibers are safe. Amorphous fibers have been used successfully in some bioimplants. Later we will discuss some amorphous fibers that should be safe and benign, but in fact seem to have a potential to be deadly. Phase chemistry and thermodynamics of polyphosphates will be discussed in more detail in Chapter 3 and their uses discussed in Chapter 8. Some amorphous phosphate-silicate fibers have been found to be very toxic when injected into rats.

1.8. Whiskers

Whiskers are a class of single-crystal fibers with remarkable properties. In this respect $[Ca(PO_3)_2]_n$ fibers could be considered whiskers. Usually whiskers are grown at high temperatures and are single crystals with small diameters near 0.5μ and lengths of 50μ or more. This makes them ideal for many reinforcing applications. Abe's group in Japan is engaged in some splendid new work in this area that is destined to be of great value.

It should be obvious that care should be exercised when handling or working with whiskers. Too little is known about most whiskers to predict their long-range impact on health. There is small doubt they should be handled with care, because most are very small and extremely refractory.

Potassium titanate fibers are similar to a whisker in many respects, but they are grown in more conventional equipment. Titanium dioxide and potassium carbonate are usually heated in a furnace to crystallize fibers that are a beautifully uniform size with diameters near 0.5μ and lengths of 50μ or more. They are very strong and make ideal reinforcing agents for plastics and similar applications. Unfortunately their safety record is not good and they are no longer manufactured in the United States. However, they were available from Japan even after manufacture was discontinued in the United States.

Franklin Fiber is based upon calcium sulfate. Rates of solution of this fiber are relatively rapid. It has a beautiful white sheen when properly manufactured and is usually of excellent diameter and length. Unfortunately this fiber is not very strong, but it is probably safe because of its high dissolution rate. But like all fibers it should be handled carefully.

1.9. Fiber Characteristics

Some general considerations of fibers of types just described should be discussed. It is often helpful to describe properties of a "perfect object" when attempting to define the properties of real systems. Fibers become a classic example of systems of too many types, too many uses, and too many variables to describe a "perfect fiber." Manufactured fabrics require long, strong, small-diameter fibers, an inch or more in length. On thc other hand, in reinforcing applications, such as reinforced extruded plastic products, a fiber of 0.5μ diameter and 50μ long may be ideal, while long fibers may not be desired.

As diameters of fibers decrease, surface-to-volume ratios S/V increase rapidly. When choosing a specific volume, V (1 cc) of a continuous cylindrical filament, it can be derived that:

$$S = 2/r \tag{1.1}$$

where S is the surface area of a fiber, ignoring ends, and r is its radius.

When ends are included, ends can be a significant percentage of surface, but only if a fiber is of larger diameter. Including ends of fibers, assuming ends are cut perpendicular to a cylindrical fiber, the equation becomes:

$$S = 2/r + (n+1)\pi r^2 \tag{1.2}$$

where n is the number of fibers per 1 cc of fiber. Maximum values of n must depend upon r, and the smaller r becomes the larger n can become. However, n is also limited by aspect ratio, if a fiber must have an aspect ratio of ten or more.

When classifying the "fineness" of fibers one of the stranger units is denier. It is weight in grams of 9,000 meters (9,846 yards) of fiber. If 9,000 m of fiber weighed

150 grams it would be 150-denier. Linton reviews a very interesting history of denier from the times of Julius Caesar.[14]

1.10. What Is a Safe Fiber?

What is a safe fiber? I must admit at once, "I do not know what a safe fiber is!" Laws of Logic cause it to be impossible to prove that any substance is safe. It is known as attempting to prove a negative. No matter how many tests to which a substance has been subjected and passed without problems, there are an infinite number of new tests a challenger can pose to insure safety. NOTHING ON EARTH CAN BE PROVED SAFE, BECAUSE NOTHING ON EARTH IS SAFE! It is easier to define an unsafe substance. It is not at all difficult to define an IDEAL safe fiber, but it is much more difficult to determine properties of a safe useful fiber. Most will agree cotton candy is a safe fiber. It has but two uses. It smells good and it tastes good. It also should be added that it is pretty to see. If adequate records were available, it is probably true that some person has choked to death on cotton candy at some time.

When fibers are small enough to be dusty they can enter alveoli of lungs. It is not difficult to understand why small needles collected in tiny "air sacks" of lungs would be very detrimental, if there were no good mechanism for our body to remove them from our lungs. These needles do not degrade, dissolve, or change form. This is the case for many small *refractory* needles. These particles are stable and resistant enough to attack that they probably will not change form for thousands of years and most will remain with hosts until all flesh disintegrates around them.

One mechanism suggested for causing fibers to exert a detrimental behavior toward a host, when inhaled or ingested, is to become lodged at a site and to remain there until surrounding tissue becomes injured and cells become transformed. Many substances considered harmful when inhaled are chemically benign. Particle size and shape rather than chemical composition seems to contribute most to problems caused by *refractory* fibers. This of course excludes particles that are innately poisonous irrespective of size or shape. All forms of dust are probably harmful when inhaled in large quantities or for extended periods in small quantities. There are accounts of cattle dying in Mississippi by suffocation by inhaling mosquitoes until their lungs became clogged with dead mosquitoes.[15]

1.11. Requirements for a "Safe" Mineral Fiber

Phosphate fibers are mineral fibers that were designed to slowly degrade when they became and remained moist. A fiber that is very quickly degraded or dissolved is of little value for most fiber uses. There are some uses, such as a soluble bandage, where it could be desirable to have a fiber quickly dissolved in water or other solvents. However, in most instances it is desirable that mineral fibers remain strong and have a long life, particularly when fibers are protected from agents known to attack fibers.

To design a fiber with an automatic safety property built into it requires that the fiber is strong and stable under conditions where it is used, but degrades when ingested, inhaled, or discarded. Surely automobile brakes that disintegrated the first time they were subjected to a wet street would be of small value, or a roof that dissolved or "biodegraded" in rain, requiring it to be replaced every two years, would be equally useless. These are areas where biodegradation is definitely an undesirable property of a product. Other applications included those where a substance biodegrades because some microorganism is ingesting it. This usually causes stenches and other problems such as discolorations, etc. and cannot be considered a desired property of a substance.

A bad example of biodegradation occurred in the early vinyl siding applications for homes. The siding concept was ideal. One could apply a siding to a home that could not be discerned from a painted frame structure without carefully inspecting the siding. It was expected that the siding should last twenty to thirty years without care or other maintenance. In some cases this was true. In other cases a glue was used to cement the vinyl to a substrate. This glue combined with the protection of vinyl made a perfect broth for fungi. In a few months the house across the street from mine was no longer brick-red, but black and red mottled, the subject of lawsuits, discontent, and maintenance intensive. One should have a difficult time convincing the owner that a biodegradable siding is environmentally friendly.

For a fiber to exhibit design safety there are several obvious properties it should possess. Naturally, a fiber must be made from substances that have a long and tested safety record. For example, a fiber can easily be made from sodium and/or calcium arsenate, but it is doubtful whether it would be considered safe. Arsenate fibers should exhibit some very desirable physical properties, but it is unlikely that they could ever be made safe when ingested. Conversely, properly manufactured and applied arsenate fibers should not be harmful. Farmers applied these compounds to soils for hundreds of years with little or no unwanted consequences, while many orientals ingested these compounds daily for their narcotic sensations when eaten.

Secondly, a fiber should not become adsorbed to membranes in a manner that it cannot be displaced and dispelled from an organism. Most fibers considered to be potentially harmful are usually chemically inert and their projected harm is expected to be physical, such as locally irritating a spot until it becomes inflamed.

Condensed phosphates are unstable with or without enzymes interacting with phosphates. Because polyphoshates are biodegradable, enzymes merely act as catalysts increasing the rates at which the polyphosphates degrade. An increase in rate is not insignificant. Often phosphates can degrade at rates one million times greater than they would without enzyme activity. In this sense polyphosphates are unique as a mineral fiber. No other mineral fibers are biodegradable. It is this property of polyphosphates that gives them added safety advantages necessary to have a safety record that could not be questioned even before the Phosphate Fibers Program was initiated. It was for this reason the program was initiated.

Polyphosphates used in this fiber research are by no means the only phosphate fibers known. Phosphate fibers have been known and utilized for more than one hundred years. In the chapters to follow details of these systems will be addressed. It will be demonstrated both theoretically and experimentally why phosphate fibers should be safe when used properly. If Phosphate Fibers are ultimately proved to be harmful to humanity, there probably will never be a mineral fiber that is safe!

1.12. The Phosphate Fiber Family

Of the fibers listed in Table 1.1 only $[Ca(PO_3)_2]_n$ and $[NaCa(PO_3)_3]_n$ were considered seriously as commercial fibers. These two seemed to have the properties that would be economical to manufacture, safe, and possessed properties that should help them to be useful as commodity materials. Late in the Phosphate Fibers Project's research phase Mr. Ngo prepared a new fiber, $[NaMg(PO_3)_3]_n$. There is no doubt that Mr. Ngo's $[NaMg(PO_3)_3]_n$ fiber is an outstanding candidate for a potential commercial product, but it was discovered too late in the research process to be given serious consideration. There is also no doubt that it would have been a prime candidate for the second generation of fibers that were under study.

The second generation of fibers such as $[Na(PO_3)]_n$, Maddrell's salt, IMP fibers, and $[NaMg(PO_3)_3]_n$ had no relationship to "asbestos," in any way. Unfortunately, company management was not aware of this phase of the project because it was new. It is also unfortunate that the project was not reviewed with the inventor before it was killed. There is small doubt in my mind that many lives unrelated to "asbestos" have been and shall continue to be lost, that could have and can be spared, had this project been continued. Work was under way to attempt to prepare safe, nonflammable infant wear, nonflammable mattress covers, and nonflammable interiors of aircraft, as starters.

These nonflammable fibers were to be made from IMP, used for years to make toothpaste and emulsified cheese, eaten by millions of people daily. There is every reason to believe that safe fabric can still be made. I did not attempt to revive the project once the announcement had been widely publicized that the Phosphate Fibers Project had been abandoned. It should have been easier to have raised Lazarus from the dead, once the announcement was made public.[16] The question from inside and outside the company that I have confronted for many years, "Tell us the TRUTH, what was really wrong with that stuff you tried to pass off on everyone? No one would kill a project with this potential, for that reason. Do you think we are stupid?"

"Brutus is an honorable man."[17]

Table 1.1. The Phosphate Fiber Family of Prepared Crystals

$[Ca(PO_3)_2]_n$	Calcium polyphosphate (calcium metaphosphate)
$[NaCa(PO_3)_3]_n$	Sodium calcium polyphosphate (sodium calcium metaphosphate)
$[LiCa(PO_3)_3]_n$	Lithium calcium polyphosphate
$[NaMg(PO_3)_3]_n$	Sodium magnesium polyphosphate
$[NaH_2(PO_3)_3]_n$	Sodium dihydrogen polyphosphate (sodium acid metaphosphate)
$[Na_3H(PO_3)_4]_n$	Trisodium hydrogen polyphosphate (sodium acid metaphosphate)
$[Na(PO_3)]_n$	Maddrell's salt, IMP (insoluble sodium metaphosphate)
$[Na(PO_3)]_n$	Sodium Kurrol's salt
$[K(PO_3)]_n$	Potassium Kurrol's salt

1.13. Guilt by Association and Tort Reform

In an industrial setting few projects have a chance of enduring without the expressed purpose of fulfilling a recognized need. To prepare an item of commerce that no one needs is folly. In my work, I attempt first to identify a need that our storehouse of structures and properties may fulfill. Then we initiate probing experiments to determine if the expected results can be obtained.

In the case of phosphate fibers the need for a safe mineral fiber was both its strength and, ultimately, its demise. Phosphate fibers were sold to management as a safe replacement for "asbestos." To the uninformed this means it is something like "asbestos," but with only some of its dangers. Management not only accepted phosphate fibers as a viable replacement for "asbestos" in many of its uses, but actively supported the project. This is not a difficult position for management because an item of commerce is a long time away and most new products fail before anyone has to make the hard decisions.

Unlike most new potential products phosphate fibers looked better and better the more we learned of them. Many potential customers who had been supplied samples were showing ever increased interest in applications ranging from brakes for automobiles to gaskets or backings for floorings. It was becoming obvious that a new product was being demanded and research was no longer a rich manager's trifle. Phosphate fibers were for real! The capacity of our small plant was beginning to be exceeded.

The Verdict

In most cases when a product is terminated, that is the end of the story and no one ever hears of it again. This was not the case with phosphate fibers. The product had become well enough known to require some kind of announcement as to why we were not to manufacture a product that had been touted as a great new hope for all humanity.

The product is a perfect example of having everything in one package: health, safety, love of humanity, country, and citizenship rolled into one. The product was a perfect vehicle as an example of a needed product the public could not have because the product was also a perfect target for lawsuits. After all, it was the new "asbestos" and who cares whether or not it is safe, as long as it can be claimed to cause all manner of illnesses never before known to science? It probably would cause "phosphotosis" and "eutrophic lung" by the time Justice and Hippocrates completed their examination.

On July 3, 1987 the St. Louis Post-Dispatch Business section carried a headline,

> Risk of Suits Kills Monsanto Fiber Project.
>
> "Monsanto Co. is ending its development of a phosphate fiber that was seen as a replacement for asbestos in some uses.
>
> A notice posted on company bulletin boards Thursday said: "The decision to terminate this program was based on the litigation climate that exists in the U.S. Regrettably this has led Monsanto to conclude that we are not prepared to accept the potential product liability risk associated with marketing this reinforcing fiber, no matter how safe it may be."

The staff writer, David Nicklaus, goes on to say,

> "The product was seen by top management to be an important one. Richard J. Mahoney, Monsanto's chief executive officer, mentioned phosphate fiber at the company's 1984 annual meeting as one of several important products coming out of Monsanto's research laboratories."

Similar stories were carried in newspapers and publications throughout the world and the product became a focus for the need for tort reform.

On May 19, 1990 the New York Times, in the Consumer's World section, carried a headline,

> A Product Dead-Ended by Liability Fears
>
> "It seemed like a dream come true: a virtual miracle fiber touted as being as tough as asbestos but completely safe and versatile enough to be used in products ranging from roofing materials to speckling compounds to brake shoes."...
>
> "Today, the Monsanto fiber, a compound known as calcium sodium metaphosphate, is a cause celebre. Vice President Dan Quayle called it the "quintessential example" of why new Federal laws are needed to limit awards in product liability suits. His comments were made in an article he wrote in the March issue of Product Safety & Liability Reporter, a regulatory newsletter.

> "There may well have been a safe effective asbestos replacement on the market and now there isn't," Richard J. Mahoney, Monsanto's chairman and chief executive said in a telephone interview."

The article tells of a test performed by toxicologists who may have known poisons, but knew nothing of polyphosphate and gelatin chemistry. It has been known for many years that polyphosphates and gelatin form coascervates.[18,19] Leather is tanned by treating the gelatin in hides with polyphosphates, to form coascervates that are more resistant than the leather itself.[20,21]

> "To prove the fiber's safety Monsanto conducted a series of animal studies beginning in 1983. (Author's comment: It is impossible to PROVE that anything is safe.) By 1986, one study, in which fibers were implanted in gelatin and imbedded in the lungs of rats, showed that the substance produced tumors at about the same rate as asbestos, according to documents that were filed with the Federal Environmental Protection Agency.
>
> Monsanto said that the test was meaningless since the gelatin coating prevented the fibers from dissolving naturally. But EPA scientists rejected that argument, concluding that in terms of the specific test, results suggested that the fiber could cause cancer. Agency officials added, however, too little is known about the substance to gauge if it poses a cancer threat to humans."

The details of this test will be discussed in Chapter 9.

> "If you can bear to hear the truth you've spoken
> Twisted by knaves to make a trap for fools,
> Or watch the things you gave your life to, broken,
> And stoop and build'em up with worn out tools;"
>
> *"If"*
> Rudyard Kipling

The tort reform story is just beginning. In the early 1970's there was very little support for the idea that energetic ignorance might do more to harm our planet than to help it. There was no lack of energy and new wisdom expended on our planet. After four and one-half billion years we are on our way to protect our planet as it has never been protected before.

Field Marshall Rommel divided his men into four classes:

Intelligent and Deliberate: These men I put on my staff, because I can trust their judgement.

Intelligent and Energetic: These men I promote to field officers, because they will defeat the enemy.

Lethargic and Dull-Witted: These men I made foot soldiers, because they will follow orders.

Energetic and Dull-Witted: These men I rid myself of, because they are a menace to everyone.

* * *

In his recent book, *Breaking the Vicious Circle: Toward Effective Risk Regulation*, Hon. Stephen Breyer points out that the risk associated with asbestos in schools is one in eleven million and that each year 2,200,000 Americans die of all causes. A small percentage of all deaths are caused by anything other than natural causes that were programmed into us at the moment of conception.

I was elated to discover Judge Breyer's book even though it was very late in the writing of this book. He has approached the message I am attempting to bring to you from an entirely divergent perspective. His development is vivid and I feel that anyone interested in public protection, environment, or government should read his book. It is an outstanding work. The fact remains, the ticking of clocks is still the single greatest hazard confronting all of us. It is time pollution.

1.14. Conclusion

In summary, the word "asbestos" has been misused. There is no substance in the universe that is "asbestos." The substances referred to as "asbestos" are truly many serpentine minerals as diverse in properties as rubies and diamonds. It is believed that some forms of serpentine mineral fibers are indeed hazardous when improperly used, but it is also believed that chrysotile is not a truly dangerous substance. This is particularly true for casual contact. The amphibole fibers are believed to be the more refractory and potentially more dangerous substances.

There is yet need for a very safe inorganic "mineral" fiber. It is also believed that the mineral fiber substitutes derived from phosphates are as safe as most substances on Earth. Although phosphate fibers are inferior to chrysotile in many of the applications for which "asbestos" was used, there are many applications in which phosphate fibers are superior to the mineral fibers. It is the author's conviction that there are theoretical reasons that nothing else exists which is safer than a condensed phosphate mineral nor can anything ever be found which is safer. Life on Earth was initiated and evolved this way requiring no input from mortals.

1.15. About the References

Throughout this book only the references that were directly used to invent and prepare phosphate fibers have been included. A literature search recovered hundreds of references that are related, but were not actually used in this work.

1.16. References

1. *Asbestos; Its Sources, Extraction, Preparation, Manufacture and Uses In Industry and Engineering,* p. 15, Becker & Haag, Berlin (1928).
2. Personal involvements of the author in some cases, many media reports in others.
3. B. A. Andrews, *Encyclopedia of Chemical Technology*, Vol. 7 (M. Grayson, ed.), p. 193, Wiley, New York (1979).
4. *Encyclopedia of Science and Technology, Vol. 5*, p. 243, McGraw-Hill, New York (1960).
5. Anon., *Chem. Eng. News* **45**, 30 (1967).
6. Accident Facts, p. 7, National Safety Council, Chicago (1989).
7. D. E. Lilienfeld, J. S. Mandel, P. Coin, and L. M. Schuman, *Br. J. Ind. Med.* **45**, 283 (1988).
8. P. Brodeur, *Expendable Americans*, Viking Press, New York (1974).
9. W. Eitel, *The Physical Chemistry of the Silicates,* p. 42, The University of Chicago Press, Chicago (1954).
10. J. Bauer, *Minerals, Rocks and Precious Stones*, Cathay Books, London (1974).
11. J. M. Hughes, H. Weill, and Y. Y. Hammad, *Br. J. Ind. Med.* **44**, 161 (1987).
12. J. E. Lockley, *Clinics in Chest Medicine* **2**, 203 (1981).
13. An interview by the author of a St. Louis gynecologist who had reported talc problems in his local news letter.
14. G. E. Linton, *Applied Basic Textiles*, p. 295, Duell, Sloan and Pearce, New York (1969).
15. Told to the author by a cattle farmer in southern Alabama.
16. *The Holy Bible*.
17. W. Shakespeare, *Julius Caesar* (1564–1616).
18. M. V. Fogle and G. Horger, United States Patent No. 3,697,437 (October 10, 1972).
19. T. Link and C. Thies, *Coulombic Interactions in Macromolecular Systems*, p. 240, ACS Symposium Series 302, Washington, DC (1986).
20. K. H. Gustavson and A. Larsson, *Acta Chem. Scand.* **5**, 1221 (1951).
21. J. R. Van Wazer, *Phosphorus and its Compounds*, p. 467, Interscience (Wiley), New York (1958).

2

The Limiting Element for Life on Earth

2.1. Primitive Earth

It is the purpose of this chapter to furnish a background for understanding the indispensable nature of inorganic phosphates in life processes. If there were no phosphates, there would be no food for our food chain. It is self-evident that phosphates have been compatible with life for hundreds of millions of years. Their compatibility with life and life nurturing qualities make them ideal substances from which to manufacture fibers and other materials of commerce. We have no reasons to fear them.

Inorganic phosphates can be trusted and they can be made to represent the balance between useful refractories and degradable nutrients. This causes them to be unique; no other inorganic substance, other than water and oxygen, is so vital in these life-giving properties at every stage of living.

Included in this chapter is a history of phosphorus on Earth with explanations and speculations of how and why the element evolved. Ore deposits from which we mine phosphates can be traced through events that give clues to where new mines for phosphates can be located for future excavations. The Tennessee deposits are found along the shoreline of ancient seas, while the western United States' deposits, high in the Rocky Mountains, were ancient seabeds. The deposits of the Kola peninsula of Russia are mostly igneous in the area near the city appropriately named Apatity. We will treat these deposits as primary.

Sedimentary phosphate ores often are rich in ancient bone fragments and must be secondary deposits. When we mine these bones and convert them into food products or fertilizers we inject ourselves into a food chain that was active 200 to 600 million years ago. Almost everything involved in biological processes has been recycled many times over. The raw materials to sustain life are simply too limited to support the biomass of Earth without using them repeatedly. If this were not true, a food chain where every living substance is devouring other living substances to survive would not be required. Everything, including you, has been here since the lithosphere cooled. There is small doubt that a part of your chemical makeup was at one time a part of George Washington or Julius Caesar. Energy alone constantly enters and leaves Earth. Over the aeons slightly more energy has been received and stored as fossil fuels, vegetation, and oxygen than has been exhausted back into the universe.

The average energy entering and leaving Earth has been in a state of dynamic equilibrium for at least 600 million years, about the same time phosphates became abundantly available. Since our only source of new energy is from the sun, there is small reason to believe that the sun's radiation has changed appreciably in the last billion years. For Earth to remain in balance, energy is irradiating back into the universe at the same rate it arrives. Obviously this process has lacked significant changes, except for short-lived minor perturbations. These upsets may have lasted long enough to have created dramatic changes in the flora and fauna of Earth, but the biomass of Earth has been nearly constant since the time phosphates first became abundantly available. Before Copernicus, Earth was the center of the universe by divine decree. Protection agencies have relieved the church of this responsibility.

While considering recycling it also should be stated that the laws of conservation of mass and energy have not been repealed. Humanity has never succeeded in creating one pound, not even one gram, of anything. We don't know how to create matter and therefore we must rely on the very natural substances available. Moreover, neither do we know how to create energy. The energy that is derived from atomic reactions as either fission or fusion is nothing more than packing fraction energy. Very small quantities of energy are released that was placed on Earth during accretion and fortunately only minute quantities of mass are converted back to energy.

2.2. Phosphate Ore Deposits

I am more familiar with the Tennessee phosphate ore deposits of the southeastern United States, having spent more than thirty years experimenting with Ilite clay tailings from these deposits.[1] The mining tailings are rich in microscopic bones. Many vertebrae of tiny organisms can be found. The creatures were probably swimming near the shoreline of eutrophic inland seas when they died for whatever reasons. The phosphate mines from which the tailings were taken occasionally contain skeletons of mastodons that died on the shorelines at a much later time. These deposits, classed as Ordovician, are about 450 million years old.[2]

One can only speculate about the conditions in the expansive inland seas that became bountiful in life and rich in phosphates with no input from humanity because there was no humanity. And how did marine bones become sources of ore in phosphate mines 7,000 feet above sea level in the rocky mountains of Idaho? (The reasons are well understood, but could it have been that pre-cave men were washing dinosaur skins in the *Surf*® and *Tide*®?) Nevertheless, these ancient treasurers testify to the importance of phosphates to all living creatures throughout Earth's history. They are truly one of modern humanity's most wonderful gifts. Without them we should all be hungry.

2.3. Phosphates in Living Systems

Reasons for phosphate's need in living systems predate life on Earth. To appreciate phosphate's significance in maintaining life requires a review of Earth's creation and the status of phosphorus on Earth throughout Earth's history. It is argued life was very sparse during Earth's first three and one half billion years. The first forms of life that have been found are nothing more than algae-type plants imbedded in silicious deposits of amorphous rock. To flourish on a maturing Earth life awaited generous supplies of phosphorus as *soluble* phosphates. It is equally true today. Food is bountiful only where phosphates are plentiful. It is well recognized that life cannot exist without phosphorus.[3]

Literature dealing with phosphorus in living systems is extensive and complex. Perhaps it will be helpful to begin with the obvious requirements for phosphates in vertebrates without becoming involved in the general biochemistry of phosphorus. To treat this very important subject superficially would likely be more harmful than helpful. Several excellent references will be cited for background and a starting place. Most textbooks on general biochemistry devote a section to phosphates. *Phosphorus Metabolism*, although old, is an excellent starting point.[4] The articles by B. Katchaman in Vol. II of *Phosphorus and Its Compounds* are still classics and deal with many aspects of phosphorus in living systems.[5] Much progress has been made since these articles were published, but the fundamentals are still more or less the same. Kulaev's book, *The Biochemistry of Polyphospates*, is excellent.[6] Kornberg's many articles are also classics.[7,8] In his more recent works the genes in DNA of *E. Coli* responsible for the formation of enzymes important in both the formation and degradation of polyphosphates have been identified.[9–11] This has profound influences on our understanding of the behavior of these systems.

It is impossible to know whether or not the properties of phosphates dictated the characteristics of life or the characteristics of life required the unique properties of phosphates for life to sustain itself. Certainly similar questions could be asked of water, oxygen, carbon, nitrogen, and all other essential components for life. The following is a discussion of focus and not of exclusion. The properties that I consider vital are listed.

1. Stability of oxidation state is a major factor. Almost all phosphorus is found in an oxidation state of +5 and it is very difficult to reduce phosphorus to

lower oxidation states. The fact that tricalcium orthophosphate, called apatite, called phosphate rock, called bone phosphate, called bone ash, must be heated above 1500 °C to reduce the phosphorus to elemental phosphorus, attests to the stability of the oxidation state of phosphorus in phosphates.

2. Solubility of calcium orthophosphate is another consideration of phosphates in living systems. The calcium orthophosphates in bones and teeth is very sparingly soluble. Conversely, many phosphates ingested are soluble. This allows them to be transported throughout a living body, yet mineralized at points where they are required.
3. Orthophosphates can be converted to polyphosphates, such as adenosine triphosphate, where they have new properties of an anionic substance with a relatively high charge. This allows the polyphosphate to sequester cations as calcium and magnesium yielding locally very high concentrations of these ions at reaction sites while the concentration of a total system can be very dilute. This can happen while the molecule is entering into reactions such as phosphorylations. These reactions are critical to metabolism and photosynthesis.
4. Hydrolytic degradation and polyphosphate formation are both vital and are both subject to enzyme interaction for control.
5. Buffering capacity and buffering range of phosphates are also critical variables. The maximum range of pH stability for phosphates is near pH 7.
6. Inorganic phosphates, both ortho- and condensed phosphates, are highly ionized.
7. There is but one natural isotope of phosphorus found in Earth's environment and it is the very stable P^{31} isotope. Whether or not this is a requirement for a building block for life is speculation, but it limits the likelihood that the element might undergo spontaneous radioactive decay in the DNA of an organism. Some forms of decay can create a new molecule, such as a sulfate that might not behave as desired in a biological system, even if the molecule withstood the impact of decay. An unstable phosphorus atom in the bridging phosphate of DNA could be disastrous for an organism.

It is reasonable to assume that just after accretion Earth was a molten mass of spherical form, and of more or less uniform surface temperature. Earth was probably enshrouded in a dense cloud cover of water vapor, while energy acquired from the sun was insignificant when compared to the energy Earth was generating. Since Earth's surface was above water's boiling temperatures, Earth's entire outgassed water supply, including all primitive oceans, was contained in these extremely dense clouds. As Earth began to cool, mostly by irradiation, some energy was lost as gases escaped the upper atmosphere.

Earth's surface eventually cooled below the boiling temperatures of water. Oceans contained in clouds of a very dense atmosphere formed rapidly as vapors condensed and rains poured as slightly acidic fresh water. It was the first of the "acid rains." This point will be revisited.

If it is accepted that the above picture is reasonable, then in the beginning phosphates were locked in igneous rock and there was insufficient *soluble* phosphate to maintain more than wisps and promises of life to come. Fresh water oceans acted as giant sinks for soluble phosphates, holding them in dilute solutions until all seas ultimately became saturated after about three and one-half billion years.

I approach this solubilization problem by making some simplifying assumption for a starting time; Earth was a molten sphere cooling to an igneous surface.[12] This condition probably never existed everywhere at once, but as a starting model it would change things very little after two or three billion years. Today, approximately one-half the phosphorus of the lithosphere (the solid crust) is in igneous deposits and approximately one half in sedimentary deposits. If the entire lithosphere was never at any time a single continuous igneous deposit, individual phosphate deposits surely were. Assume that igneous deposits of today formed in the beginning and that new igneous deposits are but a small part of total reserves of igneous phosphate.

2.4. Igneous Rocks to Sedimentary Rocks

For simplicity, igneous phosphate rock is defined as rock formed when primary primitive magma cooled. It will be assumed that most of the igneous phosphate rock of Earth has been igneous phosphate rock since the lithosphere first cooled and has not been through more than one heating and cooling cycle. This is surely not true in all cases, but is approximately correct.

Sedimentary phosphate rock is rock that has been formed from the igneous phosphate rock by weathering. The rocks may dissolve and reprecipitate, or may be ground to powders small enough to be suspended and transported, by water or wind, from its initial primitive deposit to some new deposit. Unlike the igneous deposit from which it was derived, sedimentary deposit may have been through numerous relocations over the ages.

The sedimentary phosphate rock deposits that are mined may have been in several life cycles. Unfortunately, there is as yet no reliable method to date phosphorus in the same manner that carbon and some other elements can be used for dating. As mentioned above, there is but one stable natural isotope of phosphorus.

A question arises: how long did it require for phosphates contained in igneous rock to be leached to a degree that all oceans of Earth became saturated with respect to calcium orthophosphates? It is well established that all oceans of Earth are and have been saturated with respect to calcium orthophosphates for hundreds of millions of years.

In the paragraphs to follow, the geology and geochemistry of phosphate deposits, their origins, compositions, weathering, tectonic upheavals, and the like are grossly oversimplified in an attempt to arrive at a mathematical model that can be approached simplistically. For a much more elegant description of the phosphated deposits, the first nine chapters of *Environmental Phosphorus Handbook* are highly recommended. Probably, no more authoritative nor concise presentation of this fascinating history exists in any other collection.

2.4.1. Aging and Leaching of Igneous Rocks

At equilibrium, the least soluble substance in a system that can form will precipitate. Much phosphate contained in sea water is precipitated as tricalcium orthophosphate or hydroxyl apatite, $Ca_{10}(PO_4)_6(OH)_2$, and fluorapatite, $Ca_{10}(PO_4)_6(F)_2$. Oceans' floors are covered with these deposits and are referred to as marine pellets. There are many ways in which this problem may be approached, but it is obvious that if phosphates are to be leached from igneous rocks, large boulders will leach very slowly. Smaller particles of rock caused by grinding, weathering, and aging solubilize more rapidly than larger particles. As a first approximation, rates of solubilization are proportional to fresh surfaces of solubilized rocks.

An interesting current example of the grinding and milling influences caused by glacier activity can be found in the mountain streams at the foot of Mt. Cook in New Zealand. The melting of ice from the Tasman and Richardson glaciers feeds lakes Tekapo and Pukaki at Hermitage, New Zealand. The glaciers have ground stones to a colloidal powder so fine that Lake Tekapo and Lake Pukaki are azure gray, as though the waters had been mixed with about one-tenth percent of Portland Cement. Irrigation canals many miles from the lakes continue to exhibit a milky gray color because the powdered rocks settle so slowly.

As a casual observation, it appears that the artificial mountain lakes in New Zealand, used to generate power and supply water for irrigation, are filling with glacial silts at a rapid rate. Treatment before the problem becomes acute could extend the life of these lakes indefinitely. Without proper treatment it is doubtful whether these artificial lakes will be useful for more than a few decades. The peoples of New Zealand have clearly voiced their opposition to continuing to increase the elevation of dams in order to solve electrical power problems that are certain to come. Once a lake is filled completely there is no choice but to increase the elevation of the dams, if the lake is to continue to serve a useful function.

It can be assumed that the rate of formation of new surfaces to be solubilized is proportional to the total surface of a mass. This merely means that small lumps can be crushed to finer particles more rapidly than larger boulders can be broken up, crushed, and ground to fine particles with much increased surface area. Within limits, small particles grow in surface area more rapidly than larger particles and satisfy the relation

$$dS/dt = kS \tag{2.1}$$

Upon integration,

(2.2) $$\ln S = kt$$

or

(2.3) $$S = S_0 e^{kt}$$

where S is surface, expressed in units of the original surface, S_0, at time $t_0 = 0$, and k is a proportionality constant. At time zero, S_0 equals unity. When surface layers of Earth had broken enough that the original surface area had doubled, then $S = 2$, etc. Of course S can represent the total surface of igneous phosphate on Earth. If one chooses not to use the entire Earth as a single system, surfaces of all igneous phosphate rock can be equated to unity at time zero. Either model is satisfactory since an original surface will be assigned an initial value of unity.

As more mass of igneous rock is reduced to smaller and smaller particles, the more rapidly will it be reduced to finer particles. Rates of formation of new surface are logarithmically greater as more surface forms.

2.4.2. Oceans Become Saturated with Phosphates

Next, assume that the rate of phosphate leached from surfaces is a function of surface area leached. As surfaces of a mass increase, rates of leaching increase by a factor k':

(2.4) $$dP/dt = k' S_0 e^{kt}$$

where $S_0 = 1$ and t_0 = zero. Upon substituting for S we obtain

(2.5) $$dP/dt = k' e^{kt}$$

and upon integrating

(2.6) $$P = k/k' e^{kt}$$

Now, if k is assumed to be of the same order as k', then

(2.7) $$\ln P \approx kt$$

An estimate of k can be obtained by utilizing the estimate of Horn and Adams that 22.4×10^{14} tons of phosphorus are contained in sedimentary deposits.[13] Also, assume that t_0 was 4.3×10^9 years ago. Then,

$$k = \frac{\ln 22.4 \times 10^{14}}{4.3 \times 10^9 \text{ years}} = 8.2 \times 10^{-9} \text{ year}^{-1}$$

Present quantities of phosphorus in the oceans of Earth have been estimated to be 9.8×10^{10} tons.[14] Therefore,

$$t = \frac{\ln 9.8 \times 10^{10}}{8.2 \times 10^{-9}\ \text{years}^{-1}} = 3 \times 10^{9}\ \text{years}$$

It required about three billion years for all the oceans of Earth to become saturated with phosphates for the first time. Until then oceans acted as a giant sink for phosphorus, holding it in solution too dilute to be of much use as a nutrient to life forms. When Earth's oceans became saturated, any new dissolved phosphates entering the seas caused precipitation of exactly the same amount of phosphate someplace else. This action became faster and faster until today there is a complete replacement of phosphate in oceans every 4.9×10^4 years, a very short time on a geological time scale. This estimate is based upon the estimate of McKelvey *et al.* that 2×10^6 tons of phosphorus are delivered to the seas each year, and Horn and Adams' estimate of the phosphorus in the seas.[13,14]

When oceans became saturated, large sedimentary deposits began to form more and more quickly and it was about 500×10^6 years before crustaceans and then vertebrates became plentiful. Phosphates are required for the formation of vertebrates since their framework is calcium phosphates. There is small reason to doubt that abundant life on Earth required abundant phosphates to support this life. Life was forced to await a source of phosphates. Not only are phosphates not harmful to life, phosphate fibers are one of the purer forms of phosphates known, and life cannot exist without phosphates.

When seas became saturated with phosphates, the compound that precipitated around shorelines where rivers emptied into oceans or in evaporating inland seas was hydroxyl- and fluor-apatites. Most organisms living in these seas should have migrated to shorelines or drifted there when dead. As sedimentary deposits formed rapidly, soluble phosphates became much more abundant to all forms of life.[15] Phosphates are useless to plant life unless solubilized. As discussed in detail below, this is the reason that mined phosphates are treated with sulfuric acid to make them soluble enough to act as fertilizers. Even after being acidified their value as a fertilizer is short-lived because they are precipitated by the minerals of soil in much the same way they were insolubilized in sea water.

2.4.3. Retention Times

Retention time for minerals in a body of water is the time required for current soluble salt under consideration to be removed from the water and replaced by new soluble salts. This can happen by all manner of actions from chemical precipitation, evaporation of water, removal by living organisms, and any other process that might occur.

Highly soluble salts have extremely long retention times while some salts, such as calcium orthophosphates, are very slightly soluble and have short time cycles. There is a small quantity of phosphate at equilibrium in sea water and it requires a small quantity of new salt to replace it. Therefore, if there is an adequate source of these salts, such as a sedimentary deposit with slightly acidic fresh water rains falling on them, turnover times can be as short as 50 to 100,000 years in oceans.

Retention time as used here is not to be confused with turnover times for phosphates in living organisms such as algae or bacteria. In lake waters these turnover times can be of the order of minutes. In this sense turnover time is the time for a phosphate to be ingested by one organism and then transferred to another. Retention times in oceans for a mineral to be completely replaced may be of the order of a few thousands of years to untold billions of years.

Humanity is completely helpless to change the concentration of phosphates in oceans, either up or down, intentionally or otherwise. Moreover, if all phosphates of Earth were solubilized and dumped in oceans instantaneously, the concentration of phosphates would undergo only a small change for a very short period. There is enough dissolved calcium and magnesium in sea water to precipitate very quickly all phosphate entering seas and oceans. The idea proposed by some that humanity can cause oceans of Earth to become eutrophic with respect to phosphates is embarrassing nonsense.[16]

At this writing there is a new warning being delivered by television stations that the Mississippi River is causing excessive growth of plant life in the Mississippi Delta, killing fish as a result of phosphate fertilizers. The Mississippi River is and always has been the colon of the United States. Every dead wild animal dying between the Rocky Mountains and the Allegheny Mountains washes into the Mississippi River. It has always had a very heavy loading of phosphates, just as the Nile was rich in phosphates that helped feed the fish of the Mediterranean Sea. It has been claimed that the Aswan Dam had a detrimental influence on fish yields by lowering the phosphate entering the Mediterranean Sea.

Retention times for most soluble salts are extremely long. The most abundant salt in oceans is sodium chloride. Oceans of Earth have never been saturated with this salt and it has never turned over once since Earth's beginning. This does not mean that local deposits have not formed in large quantities and that large quantities have not been removed and replaced. It merely means that oceans have not become saturated with respect to sodium chloride, potassium chloride, ammonium chloride, sodium nitrate, and other soluble salts as they have with calcium phosphates.[17]

More soluble sodium chloride can be put in oceans and it will remain dissolved, but no more soluble phosphates will remain dissolved in oceans, independently of how much is added from any source. Any new additions merely become insoluble and collect on the bottoms of oceans as pellets that have been collecting for millions of years. As mentioned, phosphates must be dissolved for plants or animals to assimilate them. This should be very gratifying for those concerned about phosphates and humanity's influence on our seas and oceans. All phosphate that can be dissolved in oceans is already dissolved, and this has been the case for more than six-hundred million years!

Even before the lithosphere cooled sufficiently, allowing oceans to form, Earth already had most of its supply of phosphorus. Small quantities of phosphorus continue to rain on Earth today from meteors and interstellar dust, but this quantity is insignificant with respect to Earth's initial store. When Earth cooled enough to form an incipient solid crust, all phosphorus and silicon had been oxidized to their highest oxidation state. Today, no vestige of elemental phosphorus or lower oxides of phosphorus exists as minerals, nor does one find any extensive deposits of silicon metal. I was told that a small natural deposit of metallic silicon exists some place on Earth, but I have been unable to verify this observation.

2.4.4. Primitive Earth's Atmosphere

Earth's atmosphere has never been reducing with respect to silicon or phosphorus. There are also reasons to question whether or not Earth's atmosphere was reducing with respect to carbon. It is well known that phosphorus, silicon, and carbon are *oxidized by water*.

$$2P + 8H_2O \rightarrow 2H_3PO_4 + 5H_2 \tag{2.8}$$

$$Si + 3H_2O \rightarrow H_2SiO_3 + 2H_2 \tag{2.9}$$

$$C + H_2O \rightarrow CO + H_2 \tag{2.10}$$

$$CO + H_2O \rightarrow CO_2 + H_2 \tag{2.11}$$

Equation (2.10) represents the well-known water–gas reaction. Much excellent work has been done in defining conditions under which graphite reacts with steam.[18] There is no reason to question that this reaction occurred on primitive Earth.

Water was surely abundant in early atmospheres and throughout Earth's history with a "cooled" lithosphere. If there were water, Earth had an oxidizing atmosphere, while hydrogen liberated from water escaped from Earth's atmosphere just as it continues to do today. Both hydrogen gas and helium, when released into Earth's atmosphere, escape from Earth never to return, because the thermal velocity of these gases is above the escape velocity of a mass of their size. They are two of a very few substances that are not automatically recycled. We have no other choice than to recycle unless garbage is to be disposed of in outer space, a highly unlikely event.

"Oxygen is the most abundant of Earth's surface elements, making up 88.8% of all waters (of Earth), 50% of rocks and soils, and one-fourth of the atmosphere, by weight, or one-fifth by volume."[19] There are two primary sources of atmospheric oxygen: (1) out-gassing during early stages of planet cooling, and (2) photosynthetic oxygen. If Earth's atmosphere were ever reducing after most out-gassing was completed, all oxygen would have been reacted. It is concluded that today's atmospheric oxygen is predominantly of photosynthetic origin. Replenishment of atmospheric oxygen is definitely a product of photosynthesis and occurs mostly as a result of sea flora.[20]

Fossil fuels and vegetation are the reduced form of a redox couple with oxygen. When considering the lithosphere, biosphere, hydrosphere, and atmosphere, Earth is not only electrically neutral but also redox coupled with sunlight keeping the Earth system charged. There is a stoichiometric quantity of oxidizing and reducing agents in the solarsphere (the outer layers of Earth that receive most of their energy directly from the sun) of Earth. The oxidizing agent is primarily atmospheric oxygen, while the reducing agent is mostly vegetation and fossil fuels. As with electrical neutrality there are local perturbations, but the solarsphere is summed to redox neutral.

Another way of considering this same idea is to note that, without sunlight, the free energy of the biosphere would soon approach zero. Eventually all atmospheric oxygen would react with fossil fuels and other reduced substances such as metals or elemental sulfur and sulfides. Since Earth's surface is redox neutral, in the same sense that it is electrically neutral, if all oxygen of the atmosphere were allowed to react, all fossil fuels would be oxidized back to their original state as carbon dioxide and water, the reversal of photosynthesis.

Obviously Earth's core and deeper mantle are in a reduced state, but they are protected from atmospheric oxygen by a semistable lithosphere. It is this semistable portion of lithosphere that protects fossil fuels from being oxidized by atmospheric oxygen.

2.5. Phosphorus in Photosynthesis

In photosynthesis,

$$CO_2 + H_2O \underset{\leftarrow \text{Combustion}}{\overset{\text{Sunlight}\rightarrow}{\leftrightarrow}} \underset{\text{Fuel}}{\{CH_2O\}} + O_2 \qquad (2.12)$$

Reactions represented by equation (2.12) store energy from sunlight and recycle all reduced forms of matter to keep a mobile equilibrium on Earth's surface. Phosphates do not participate directly in an oxidation reduction couple as an oxidizing or reducing agent, but they are extremely important in stabilizing this chemistry. Because phosphates are in their highest oxidative energy state and are not strong oxidizing agents, they can participate in biological reactions without destroying fragile biochemicals by rapid oxidation.

Similar arguments to those above could be made about silicates, except that while phosphates are reasonably stable toward polymerization–depolymerization reactions, soluble silicates are extremely pH-sensitive, polymerizing under acidic conditions and depolymerizing under basic conditions. It is the controlled pH sensitivity that allows phosphates to participate as an energy reserve in molecules such as adenosine tripolyphosphate, ATP. It is the basis for phosphates being one of the more biofriendly substances on Earth and ideal substances from which to prepare commercial goods contacting human environments.

Evolutionary chemists have been faced with a problem referred to as the Phosphorus Enigma. Where did early life forms find enough soluble phosphates for life to become plentiful? The answer seems to be that they did not. Although there are scattered life forms that existed between three and four billion years ago, there is no evidence that life was plentiful on primitive Earth.

2.6. Algae's Role in Controlling Acid Rain

It is believed that oceans of Earth formed rapidly after the lithosphere's temperature fell below 100 °C, perhaps requiring no more than three hundred million years. Oceans were probably acidic, because most secondary atmospheric gases derived from sulfur, carbon, nitrogen, etc. are anhydrides of acids. This is also the reason that rains from the beginning of Earth have always been acidic. A basic rain would be a strange rain indeed. It is aquatic photosynthesis that keeps pH of the waters of Earth controlled. Aquatic growths, such as algae, act as a natural buffering system producing oxygen and neutralizing acids.

$$\underset{\text{CarbonicAcid}}{\overset{\text{Acidic}}{CO_3^{=} + 2H^+}} \quad \underset{\text{ATP}}{\overset{\text{Sunlight}}{\Rightarrow}} \quad \underset{\text{Algae}}{\overset{\text{Basic}}{\{CH_2O\}}} + O_2 \qquad (2.13)$$

Not only does aquatic algal growth help keep the pH of natural bodies controlled, but it should be noted that oxygen utilized by fish and other aquatic faunae is derived from the water that they inhabit and not from air above the surface of this water. This is to say that the flow of oxygen in the atmosphere is from the hydrosphere to the atmosphere, and not the other way around. Aquatic growth can cause some fish to die in the fall, but it is the same aquatic growth that allows them to live under water throughout the rest of the year. Moreover, oxygenation of water by pumping air through it is successful only on small bodies of water. If water depended upon air for its source of oxygen, all aquatic life would soon die. It can be noted that fish living in lakes rich in vegetation survive when their lake freezes over, but fish living in shallow, very clean lakes usually suffocate when their lake freezes.[21]

2.7. Life and Fibers

The biochemistry of phosphorus and phosphates is much too involved to be considered seriously in this work. Only very general concepts required to understand why orthophosphates and condensed phosphates are ideal building materials, when safety is a primary issue, will be considered. Both orthophosphates and condensed phosphates are so very much a part of any living organism; from bacteria to red woods or ameba to whales, phosphates are required for life. To reinforce what was stated above with respect to vertebrates in particular, it may be stated without contradiction, *vertebrates are phosphates*! The skeletons and teeth, if they have any, of all vertebrates are orthophosphates, classed as calcium apatites, $Ca_5(PO_4)_3OH$.

Teeth are also orthophosphates, hydroxyapatites, and fluorapatites. Fluorapatite is much harder than hydroxy apatite and it is for this reason that enamel of teeth is harder than dentin. It is also the reason fluoride toothpaste and fluorinated drinking water have had a dramatic influence on improved dental health of young Americans compared to all previous generations. It must be mentioned that fluorapatite is more brittle than hydroxy apatite and older teeth may be somewhat easier to chip as a result of fluoride treatment. The mere fact that there are billions of older teeth to chip, still working well in mouths in which they grew, is a strong endorsement of fluoride treatments. Before fluorides were introduced into drinking water and toothpaste very few older people retained their own teeth.

2.8. DNA Is a Phosphate Fiber

Living organisms are dependent upon phosphates from the moment of conception. Structural strength of strands of DNA in both sperm and egg depend upon bridges of orthophosphates to hold a deoxyribose-phosphate "fiber" together during a complex entwinement of life and chemical structure. It is an orthophosphate that is midwife to conception of life, and a polyphosphate that is nursemaid to photochemical reactions supplying food and oxygen for all new life.

```
      |                              |
Deoxyribose---ade*thy---Deoxyribose
     PO4                          PO4
Deoxyribose---cyt*gua---Deoxyribose
     PO4                          PO4
Deoxyribose---gua*cyt---Deoxyribose
     PO4                          PO4
Deoxyribose---thy*ade---Deoxyribose
     PO4                          PO4
      |                              |
```

ade = adenine; cyt = cytosine; gua = guanine; thy = thymine

There is small doubt that the only sin phosphates are guilty of is to bring life to planet Earth, but there are many other reasons why they are so well suited as building blocks of new substances, such as phosphate fibers. Above all, condensed phosphates are unstable in the presence of water. It is often stated that they are thermodynamically unstable in the presence of moisture. This means that there are good scientific reasons why they are unstable. Perhaps the best and easiest proof that they are unstable is the fact that condensed phosphates are totally unknown in the mineral kingdom. There are no mineral deposits of condensed phosphates. When they are rarely found, it is usually in guano deposit, derived from a life source of relatively recent origin. Only the dimer, pyrophosphate, has ever been reported and this deposit cannot be relocated if it ever existed. No higher condensed phosphates have ever been reported as minerals because they should have weathered back to orthophosphates.

We have considered a phosphate, ortho- or condensed, to be more or less the same. This is surely not true when considering their role in life processes. It is their unique properties that give condensed phosphates their outstanding safety properties. Not only will a structure such as a fiber, composed of condensed phosphates, fall apart to become a nutrient, but enzymes in living organisms will cause it to fall apart more than one million times as rapidly as they would in sterile systems. Condensed phosphates are the ultimate in a biodegradable substance.[22] Moreover, they fall apart to a beneficial orthophosphate that fosters and nurtures life.

In other sections considerable effort has been expended discussing how molecular structure is related to physical morphology of bodies composed of these molecules (see Chapter 4). As molecules, from which a structure is fabricated, decompose, the structure decomposes. It can be compared to a metal bridge that is allowed to rust away. A highway bridge will rust away as the steel in the bridge decomposes to rust even when the bridge is subjected to fresh water only. But, like phosphate fibers subjected to enzymes, the bridge will rust more rapidly when subjected to sea water unless it is well protected.

2.9. The Classification of Phosphates

J. J. Berzelius was first to class phosphates by their $M_2O–P_2O_5$ ratio while Thomas Graham was first to classify phosphates as ortho-, pyro-, and metaphosphates in 1833.[23] He recognized that they suffered from hydrolytic degradation or hydrolysis back to orthophosphates. For example:

$$(2.14)\quad \begin{array}{c} \mathrm{O\;O\;O\;O\;O\;O} \\ \mathrm{XOPOPOPOPOPOPOX} \\ \mathrm{O\;O\;O\;O\;O\;O} \\ \mathrm{X\;X\;X\;X\;X\;X} \end{array} + \mathrm{HOH} \Longrightarrow \begin{array}{c} \mathrm{O\;O\;O\;O\;O} \\ \mathrm{XOPOPOPOPOPOH} \\ \mathrm{O\;O\;O\;O\;O} \\ \mathrm{X\;X\;X\;X\;X} \end{array} + \begin{array}{c} \mathrm{O} \\ \mathrm{XOPOH} \\ \mathrm{O} \\ \mathrm{X} \end{array}$$

until there is no condensed phosphate remaining. It was 1941 before it was definitively demonstrated by Ingelman and Malmgren that enzymes remarkably accelerated hydrolytic degradation of inorganic phosphates.[22] Interestingly enough they employed the very long chain polyphosphate, potassium Kurrol's salt, to demonstrate their findings. Potassium Kurrol's has all outward appearances of a water-soluble form of "asbestos," as has been noted many times by other authors.

2.10. ATP

Some more active phosphates in any living organism are ATP (adenosine triphosphate), RNA (ribose nucleic acid), and DNA (deoxyribose nucleic acid). Where orthophosphate in DNA and RNA can be considered functionally structural, as a sugar bridge in DNA and RNA, such is not the case for the tripolyphosphate moiety in adenosine triphosphate. To demonstrate adequately the activity of adenosine triphosphate it is necessary to consider the structural formula of this molecule (Figure 2.1).

It should be noted that the tripolyphosphate moiety of adenosine tripolyphosphate is the same molecule that created so much controversy about eutrophication of fresh waters. It can easily be demonstrated that solubility of phosphates in fresh water containing hardness is very low. Large bodies of fresh water can become saturated with respect to apatites in much the same way sea water is saturated. Small farm ponds are yet another consideration. Nutrients can be added to small ponds more rapidly than their concentrations can be controlled by precipitation.

In the past terminal POP linkages in ATP were referred to as high-energy linkages. To be sure, high energy in this sense is relative when compared to a pyrophosphate linkage of adenosine diphosphate and there is very little difference in the energy terminal linkages of ATP and ADP.[24] As early as 1960, primarily through the efforts of Van Wazer it was shown decisively that the term high energy was incorrect.[25] Hydrolytically active linkages is a better name for terminal linkages of polyphosphates with larger pseudo-first-order rate constants. While reviewing

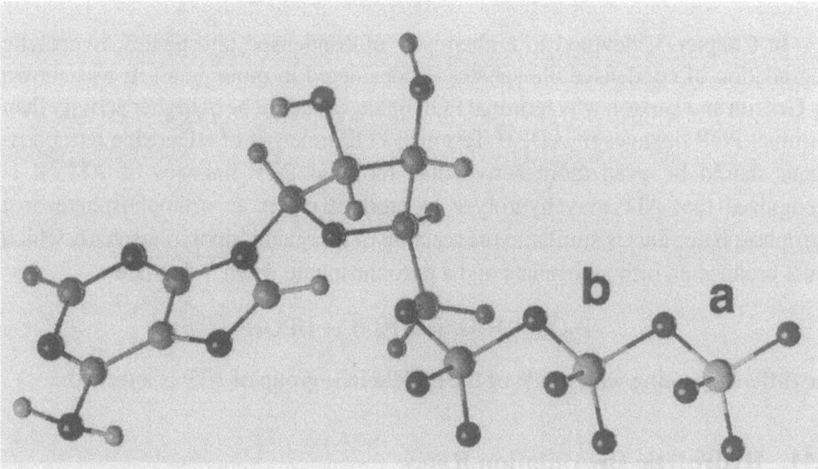

Figure 2.1. Structure of adenosine triphosphate, ATP. The POP linkages of the tripolyphosphate are marked a and b. (Molecular model by Dora M. Schnur.)

this publication at the request of the author A. Kornberg objected to the term "high energy."[26] He correctly pointed out that in enzymatic systems

$$2ADP \Leftrightarrow AMP + ATP$$

No similar reaction has ever been observed with inorganic polyphosphates except in extremely concentrated solutions at higher temperatures.[25] In the cited case the following reaction was observed:

$$2NaH_2PO_4 \Leftrightarrow Na_2H_2P_2O_7 + H_2O$$

Of course this is the way polyphosphates are manufactured.

It is known that a solid orthophosphate passes through an amorphous phase transition when being converted to a polyphosphate. It has been emphasized in the chapter devoted to phase chemistry that a melt or a solid glassy phosphate is a solution. In the case of a dissolved salt a reaction is seldom considered to be a phase transition, because there is no change of phase.

It was also shown that in enzyme-controlled systems

$$n\ ATP \Leftrightarrow \text{poly}\ P_n + n\ ADP$$

has an equilibrium constant near unity. No similar reaction occurs in aqueous media for inorganic polyphosphate, because the equilibrium constants are so small as to be practically meaningless. To be sure, if the concentrations are high enough and the temperatures are high enough this is again the way inorganic polyphosphates are manufactured.

$$n\ NaH_2PO_4 \Rightarrow Na_{n+2}P_nO_{3n+1} + nH_2O$$

In Chapter 3, devoted to a chemistry of condensed phosphates, hydrolytic degradation of condensed phosphates is considered in more detail. It was shown by Griffith and Buxton why terminal POP linkages should be of higher activity than terminal POP linkages of ADP.[27] Terminal POP linkages of adenosine tetraphosphate should be even more active than terminal POP linkages of ATP. It is recognized that ATP may hydrolyze to produce either an orthophosphate or a pyrophosphate. This is similar to the reaction of inorganic tripolyphosphate, which must produce an orthophosphate and a pyrophosphate when it degrades,

$$P_3O_{10}^{-5} + H_2O \Rightarrow HP_2O_7^{-3} + HPO_4^{-2}$$

the difference being which side of the middle PO_3 group of ATP is attacked.

2.11. Hydrolytic Degradation Rates

In the literature most rate constants are reported as first-order constants. These are pseudo-first-order rate constants, because the concentration of water is so high that its activity can be considered unity and the water is ignored. When the solutions become very concentrated this is no longer the case, and water must be considered as a reactant. Rates of nonenzymatic hydrolytic degradation of terminal linkage of tripolyphosphate are greater than rates of pyrophosphates under similar conditions. It has never been demonstrated satisfactorily that internal linkages of inorganic polyphosphates degrade except by the elimination of trimetaphosphate rings. In more than forty years internal POP degradation of a polyphosphate has never been observed. Terminal linkages become more and more active until a polyphosphate molecule with about ten POP linkages is degraded.

It has long been recognized that polyphosphates degrade by two mechanisms. The terminal group can be clipped from a molecule, but when polyphosphates contain four or more phosphorus atoms, a ring trimetaphosphate may form as well. The tetrapolyphosphate is the smallest molecule capable of undergoing this degradation, but condensed phosphates with four or more phosphorus atoms may be degraded relatively rapidly by this route.[28]

Aside from involvements of phosphorus as phosphates in the biochemistry of living systems, perhaps a second approach is to consider vital uses of phosphates in hundreds of products used daily, throughout most societies on Earth. Toy and Walsh's exceptional book, *Phosphorus Chemistry in Everyday Living*, gives an interesting approach to all forms of phosphorus applications, including DTA, RNA, and ATP. Their approach is directed more toward the usefulness of these compounds rather than their safety when in contact with live forms, but it is a simplified, excellent coverage that will not be repeated here.[29]

Since phosphate fibers are exclusively in the dominion of inorganic chemistry and therefore the mineral kingdom, these areas will receive our major attention. As a safety issue, inorganic phosphate's use in foods, fertilizers, cosmetics, and pharmaceuticals will be considered. Returning to the fact that bones when dried are a calcium orthophosphate, hydroxyl apatite, approximately $Ca_5(PO_4)_3OH$, the alliance of phosphates with life is again established. Most phosphates used in products mentioned above are from this source in one form or another. When a mummy's tomb is opened, about all that is left are phosphates he ate during his lifetime. Phosphates of his bones assisted him in his quest for immortality.

There is no doubt that much phosphate rock that is mined for fertilizer applications is used for modern humanity to inject itself into a food chain that is sixty million years or more old. Moreover, as Earth's population becomes larger our dependence on this ancient food chain grows more and more critical. It has been estimated that as much as one-quarter of the people should be condemned to starvation within four years, if we discontinued mining phosphates and attempted to live on organic phosphates alone.

2.12. Abiotic Polyphosphates

It is necessary, but not sufficient, for an abundant supply of phosphate to exist in order to support life. Condensed phosphates of Earth are formed as a result of photosynthesis and in life forms. Before there was adequate life to build new supplies of condensed phosphates and therefore POP linkages, it was required that an adequate supply of POP linkages should form by some mechanism. There are any number of ways that small quantities of POP's formed to supply initial "seeds" for life's propagation.

There are such simplistic mechanisms as lightning striking an outcropping of calcium orthophosphate that may have contained some sand or other form of SiO_2 embedded in the outcropping. To make the equation simple, we will write the apatite as tricalcium phosphate and the metasilicates in their simplest form.

(2.15) $$2\,Ca_3(PO_4)_2 + 6\,SiO_2 \xRightarrow{\text{Lightning}} P_4O_{10} + 6\,CaSiO_3$$

(2.16) $$P_4O_{10} + 2H_2O \Rightarrow H_4P_4O_{12}$$

(2.17) $$\begin{array}{c}\text{O O}\\ \text{HOPOPOH}\\ \text{O O}\\ \text{HOPOPOH}\\ \text{O O}\end{array} + H_2O \Longrightarrow \begin{array}{c}\text{O O O O}\\ \text{HOPOPOPOPOH}\\ \text{O O O O}\\ \text{H H H H}\end{array}$$

Therefore, calcium phosphate is converted to phosphorus pentoxide, P_4O_{10}, plus some form of calcium silicate. Phosphorus pentoxide reacts with water to form, first, tetrametaphosphoric acid that in turn reacts with more water to yield tetrapolyphosphoric acid. There are four POP linkages in tetra*meta*phosphoric acid and two end-group active bonds in tetra*poly*phosphoric acid. These reactions can be easily demonstrated in a laboratory. There was at one time a commercial process based upon the second and third reactions. Victor Chemical Company produced a sodium tetrametaphosphate sold under the trade name, Cyclophos. They made tetrametaphosphoric acid by mixing P_2O_5 with ice and then neutralizing this product with sodium carbonate.

Another very likely way of producing POP linkages is based upon reduction of iron in a ferric orthophosphate. There are many reducing agents that could have worked, but H_2S is a likely candidate. It will be seen in Chapter 3 that much phosphate chemistry is dictated by M_2O–P_2O_5 ratios. A ferric ion, Fe^{+3}, is equivalent to 3 Ms while a ferrous ion, Fe^{+2}, is equivalent to 2 Ms.

(2.18) $$2\,FePO_4 + 3\,H_2S \Longrightarrow \begin{array}{c}\text{O O}\\ \text{HOPOPOH}\\ \text{O O}\\ \text{H H}\end{array} + 2\,FeS + H_2O + S$$

Action in the above reaction was to form initially ferrous pyrophosphate, and then to precipitate iron as ferrous sulfide, leaving pyrophosphoric acid in solution. This conversion of either an iron or copper phosphate to a soluble form is often done by treating metal phosphates with either hydrogen sulfide or ammonium sulfide. Heating pyrophosphoric acid to temperatures around 200 °C can convert it to a long-chain polyphosphate. A POP linkage in pyrophosphate is equivalent to a POP linkage in adenosine diphosphate (ADP), but is not as active a phosphorylating agent as adenosine triphosphate (ATP).

Other potential sources of condensed phosphate from abiotic systems could come from dehydrating agents. Urea is used currently to dehydrate orthophosphates to condensed phosphates in some industrial processes. It is not clear where urea might have been generated in abiotic systems, but it could have been prepared chemically.

$$nH_3PO_4 + nNH_2CONH_2 \Rightarrow (HPO_3)_n + nCO_2 + 2nNH_3 \quad (2.19)$$

Volcanic action could have also contributed POP linkages to early environments. Some pegmatites show compounds that may have been condensed phosphates for short periods, but there are many possibilities.

The point of merit is that it was required that POP linkages come from some source before life could propagate on Earth. Once life was functional, polyphosphates are a byproduct of life, but before there was life some chemical source was required. Several demonstration models have been proposed, but there are many more possible ways that small quantities could have been available at isolated spots on primitive Earth. It is reasonably certain that condensed phosphates were not generally distributed and that they had a relatively short life in any environments where life could exist.[30]

2.13. Fertilizers

Fertilizers probably have been used in one form or another for as long as humanity has farmed as an organized and planned activity. It was surely obvious to members of even the most primitive societies that most waste products, animal or vegetable, caused plants to flourish. A next step was to add intentionally these ingredients to soils before planting. Experimentation soon showed which substances yielded better results. It was learned that one requirement was a source of bones if the crop was to yield good results. A next step was to learn that acidic bones gave better results than untreated bones. A significant observation was that phosphate rock, much of which can be bones, could be acidified to make as good a fertilizer as bones. In 1867 the first United States phosphates mines were opened in South Carolina for the expressed purpose of preparing fertilizers from ancient deposits. Canadian mines were in operation as early as 1862.

The fertilizer industry continues to improve its products to give higher and higher yields of nutrient to plants. Triple Super Phosphate fertilizer is made by first treating phosphate ore with sulfuric acid to make a crude form of phosphoric acid that is often referred to as green acid, because it is colored green and not because it has not been cured or ripened. The green acid is also called wet acid, to differentiate it from phosphoric acid made from elemental phosphorus. Phosphoric acid produced from elemental phosphorus is the purest of acids and it is called furnace acid.

Rather than treating phosphate ore with sulfuric acid as was done for many years to make Super Phosphate fertilizers, Triple Super Phosphate fertilizer is made by treating phosphate ore with green phosphoric acid. The fertilizer is a crude form of monocalcium orthophosphate, $Ca(H_2PO_4)_2$. An added value of this product is a larger quantity of phosphate contained in a pound of fertilizer and its availability by being highly soluble when it is first applied. As time passes it too becomes insolubilized as mentioned above. All phosphates remain soluble and available as nutrients for a short time on farms or in aquatic environments.

2.14. Fish Farms

"Ohna Phosphor Kein Leben." Ann.
Ohna Phosphor Kein Alge.
Ohna Alge Kein Fisch.

To which was added many years ago:

"Ohna Phosphor Kein Gedanke."

Many semiscientific articles have been written about the bad influence of phosphates on fish life. Most writers who approached this subject insisted that 10 parts per million of phosphorus in water is bad for fish because it causes algae to grow. Fish are similar to all other living organisms. Fish must eat to live and they cannot survive in beautiful sterile water. In addition to eating each other, they eat bugs, flies, worms, and vegetation. Vegetation furnishes not only food, but oxygen for them to breathe.

In recent years fish farming has been imported to the United States from China and other parts of the world. Catfish have been a major protein crop for fish farmers. To grow fish one does not require sterile, clean, pure, transparent water. Fish farm pond water must be fertilized and one of the primary ingredients used in large quantities is Triple Super Phosphate. Canadians have been quietly fertilizing their sport fishing lakes for many years to improve fish production, while banning phosphate detergents. Some similar feeding experiments have been attempted at

Lake Mead in the United States. In Lake Mead the waters are so pure and devoid of phosphates that very few fish can survive because they have nothing to eat.

2.15. The Limiting Element

The concept of a limiting element to support life in an aquatic environment was popularized during the 1960s and 1970s. Much elegance was unfurled depicting which element controlled life's development in an aqueous environment. It was believed that if a required element of life was depleted from lake water then the quantity of life in the water could be controlled. Some insisted that phosphorus was the limiting element, in some bodies of water, while others stated that it was nitrogen in other bodies. It is obvious that if phosphate fertilizers are limited below the starvation value, crops cannot grow on farms or in lakes. This is exactly what happened to the farms along the Mississippi River a few years after anhydrous ammonia was first introduced as a fertilizer. At first the nitrogen produced bumper crops, until all available phosphate was depleted. Then crops failed until phosphate fertilizers were again added to the farms. Phosphorus was the limiting element.

This limiting element concept can be carried a step further while considering that phosphorus may well be a limiting, but a plentiful element for life on our planet. Phosphorus is plentiful, because it is eleventh in abundance in Earth's crust. It is limiting because most phosphates found in the environment are so sparingly soluble, very little of it is ever dissolved.

Some lakes are classed as nitrogen limited while others are classed as phosphorus limited. There is much less phosphorus than there is nitrogen in our solarsphere. Nitrogen gas, like that found in our atmosphere, is so inert that it is controlled by its chemistry. Nitrogen is unaccessible except when "fixed" as a soluble nitrogen compound such as ammonium nitrate or urea.

There is more than twice as much phosphorus as there is sulfur on the surface of Earth.[31] But there is more than ten thousand times as much sulfur as phosphorus dissolved in sea water![32] Data are too variable to quote a number for this ratio in fresh water, but it is probably similar.

A limiting element concept has always been a bit fuzzy. Most elements that contribute to life can become very limiting, if they are in short supply. It should make little difference how much phosphorus or nitrogen might be dissolved in a body of water if there were no carbon in a form that could supply a food chain by photosynthesis. If everything else required for life is present, the limiting element idea is: Which element, if controlled in quantity, should have the greatest impact controlling the quantity of life? In aquatic environments it is possible to control an algal food chain base, without influencing other remaining environmental life. This idea was untried under real world conditions. It models well in computer programs and had some limited success in smaller bodies of water.

There is very little evidence that removal of phosphorus from household products has done much, either good or bad. Recently I attempted to revisit the phosphate question. A very involved literature search revealed that almost no follow-through work has been done since phosphates were removed from detergents. There is little tangible evidence that soluble phosphate concentrations have been lowered or raised in any major body of water. Since almost no studies have been made in the last decade, it is impossible to access a change, if any. The few studies that have been made yield no definitive answers of improvements.

It was stated definitively that the phosphorus concentration in the lakes at Madison, Wisconsin has not changed significantly in the last one hundred years. It was long ago that Dade County, Florida recognized that their ban on phosphate detergents accomplished nothing and lifted the ban in favor of superior detergents, but too late; phosphate detergents were dead.*

It is not at all surprising that those who were so adamant that phosphates were destroying our aquatic environment would not continue their studies to see if bans had actually changed anything in the concentration of phosphates in our aquatic environment. Kinetics of phosphate precipitations cause it to be expected that there would be but small change, if any, in phosphorus concentration in our aquatic environment as a result of bans. There is no doubt that removal of untreated sewage and other garbage from our aquatic environment has had a very favorable influence on our environment.

2.16. The Death of Tennessee's Phosphorus Industry

Earlier in this chapter Dr. Emigh is cited for several of the data presented. In his chapter, page 105, in *Environmental Phosphorus Handbook*, referring to Columbia, Tennessee he states: "Up until the mid-1930s most of the phosphate was mined for the fertilizer industry but since then it is used locally in electric furnaces producing phosphorus. The area is the largest phosphorus producer in the world."

The above passage was written a mere twenty years ago. As a direct result of environmental concerns and misunderstandings this industry is effete, dead, and not one pound of phosphorus has been produced in the Southeastern United States in almost five years. The entire United States has but six active furnaces today. It is reported that China is building *SEVENTY* new furnaces![†] Most of the basic industries in the United States are endangered and may well suffer the same fate. Have we reached the status that we are afraid we may soil our hands if we work? Columbia, Tennessee is not a rare, isolated example. No one seems to notice. No

*Dade County, Florida was one of the first communities to ban phosphate detergents and, after several years without phosphate detergents, was the first to lift the ban.

†A personal report by one very knowledgeable of activities in China, but who must not be named.

one seems to care. The employment is gone and this part of our natural resources is no longer resources, but dead areas of land that serve no one.

2.17. Conclusion

It was desired that this chapter would establish the strong alliance of life and phosphates, and a new appreciation for the need humanity has for phosphates. Phosphates have been portrayed as though they are an evil poisonous substance by many groups possessing very little knowledge. Each new attack has been more irrational than the preceding attacks, while new evidence that we overreacted to the initial criticism is ignored. Phosphates are our friends and we need them if we are to survive.

2.18. References

1. E. J. Griffith, R. J. Brooks, and R. A. Russell, *Trans. Soc. Metal. Exploration* **290**, 1862 (1993).
2. G. D. Emigh in *Environmental Phosphorus Handbook* (E. J. Griffith, A. Beeton, J. M. Spencer, and D. T. Mitchell, eds.), p. 105, Wiley, New York (1973).
3. C. N. Sawyer in *Environmental Phosphorus Handbook* (E. J. Griffith, A. Beeton, J. M. Spencer, and D. T. Mitchell, eds.), p. 633, Wiley, New York (1973).
4. W. D. McElroy and P. Glass (eds.), *Phosphorus Metabolism*, Vols. I and II, The Johns Hopkins University Press, Baltimore (1952).
5. B. J. Katchman, in *Phosphorus and Its Compounds* (J. R. Van Wazer, ed.), p. 1345, Interscience, New York (1958).
6. I. S. Kulaev, *The Biochemistry of Polyphosphates*, Wiley, New York (1979).
7. A. Kornberg and S. R. Kornberg, *Biochim. Biophys. Acta* **20**, 215 (1956).
8. A. Kornberg, I. R. Lehman, M. J. Bessman, and E. I. Simms, *Biochim. Biophys. Acta* **21**, 197 (1956).
9. A. Kornberg, *Protein Science* **2**, 132 (1993).
10. M. Aklyama, E. Cooke, and A. Kornberg, *J. Biol. Chem.* **268**, 633 (1993).
11. E. Cooke, M. Akluama, N. N. Rao, and A. Kornberg, *J. Biol. Chem.* **269**, 6290 (1994).
12. E. J. Griffith, C. Ponnamperumam, and N. W. Gabel, 'Phosphorus, A Key to Life on the Primitive Earth,' *Origins Of Life*, **8**, 71 (1978).
13. M. K. Horn and J. A. Adams, *Geochim. Cosmchim. Acta* **30**, 279 (1966).
14. V. E. McKelvey, R. W. Swanson, and R. P. Sheldon, *19th International Geological Conference*, Algiers (1952).
15. H. V. Sverdrup, M. W. Johnson, and R. H. Fleming, *The Oceans, Their Physics, Chemistry, and General Biology*, Prentice-Hall, New Jersey (1942).
16. E. J. Griffith, *Phosphorus in the Environment: Its Chemistry and Biochemistry*, p. 75, Ciba Foundation Symposium, Elsevier, North-Holland (1978).
17. F. W. Clarke, *The Data of Geochemistry*, U.S. Geological Survey Bulletin 770 (1924).
18. J. B. Lewis, in *Modern Aspects of Graphite Technology* (L. C. F. Blackman, ed.), p. 168, Academic Press, New York.
19. N. A. Lange, *Handbook of Chemistry*, 5th edn., p. 71, Handbook Publishers, Inc., Sandusky, Ohio (1944).
20. R. Monastersky, *Science News* **142**, 412 (1992).
21. Personal observation, Creve Coeur Lake, St. Louis, Missouri.

22. B. Ingelman and H. Malmgren, *Acta Chem. Scand.* **1**, 422 (1945).
23. J. J. Berzelius, *Ann. Physik* **53**, 393 (1816); T. Graham, *Phil. Trans.* **123**, 253 (1833).
24. B. J. Katchman, in *Phosphorus and Its Compounds II* (J. R. Van Wazer, ed.), p. 1391, Interscience, New York (1961).
25. C. D. Schmulbach, J. R. Van Wazer, and R. R. Irani, *J. Am. Chem. Soc.* **81**, 6347 (1960).
26. A. Kornberg, Private communication.
27. E. J. Griffith and R. L. Buxton, *J. Am. Chem. Soc.* **89**, 2884 (1967).
28. E. J. Griffith, *J. Pure Appl. Chem.* **44**, 173 (1975).
29. A. D. F. Toy and E. N. Walsh, *Phosphorus Chemistry in Everyday Living*, American Chemical Society, Washington, DC (1987).
30. E. J. Griffith, in *Chemical Evolution of Early Precambrian* (C. Ponnamperuma, ed.), p. 61, Academic Press, New York (1977).
31. C. D. Hodgman, *Handbook of Chemistry and Physics 31*, p. 2678, Chemical Rubber Publ. Co., Cleveland, Ohio (1949).
32. B. H. Ketchum, in *Encyclopedia of Science and Technology*, Vol. 12 (W. H. Crouse, ed.), p. 108, McGraw-Hill, New York (1960).

3

General Phosphate Chemistry (As Applied to Fibers)

General inorganic phosphate chemistry is reviewed in this chapter. Both theory and practice are covered in some detail as a background for controlling both properties and formation of condensed phosphates. Particular attention will be directed toward the necessary constraints required to grow very long polyphosphate anions resulting in long strong phosphate fibers. Much of the work to follow will refer to this chapter. The old established chemistry and some very recent chemistry that is yet too new to judge its long term value is presented.

3.1. Polyphosphate Chain-Length Control

In the example below MOM is a chain terminator. The terminator or breaker can be any metal oxide or impurity such as HOH, CaO, NaONa, etc. In the example a polyphosphate with a chain length of nine was broken into a polyphosphate of chain length four and a polyphosphate of chain length five. The chain length of this mixture is then $n = 4.5$. The M_2O–P_2O_5 ratio of the nonapolyphosphate was $R = 11/9$ before the molecule of MOM was added. After the molecule of MOM was added $R = 13/9$. This is a very simple starting place to understand the following sections.

```
   O O O O O O O O O
MOPOPOPOPOPOPOPOPOPOM + MOM ===>
   O O O O O O O O O
   M M M M M M M M M

                  O O O O         O O O O O
               MOPOPOPOPOM + MOPOPOPOPOPOM
                  O O O O         O O O O O
                  M M M M         M M M M M
```

The above reaction takes place in reorganizing melts, but it is not observable if MOM is HOH in dilute aqueous solutions.

An attempt has been made to minimize mathematical expressions presented in this work. Nevertheless, some arithmetic is necessary to understand phosphate chemistry as practiced in preparations of phosphate fibers. As a result of Berzelius notations, both silicate and phosphate chemistries are based, to a large extent, upon mixed oxides and ratios of these oxides. In phosphate chemistry R represents a molar ratio:

(3.1) $$R = \frac{M_2O}{P_2O_5} = \frac{M}{P}$$

where M is one equivalent of either metal ion or organic moiety. It is often more convenient to utilize an M/P ratio in calculations. In all cases M or P are gram atomic weights or moles.[1]

To illustrate uses of R via another simple example, tetrasodium pyrophosphate, $Na_4P_2O_7$, may be written as $2Na_2O \cdot P_2O_5$, or

(3.2) $$R = \frac{Na_2O}{P_2O_5} = \frac{2}{1} = 2.0$$

Average chain length, n, may be calculated from a knowledge of R:

(3.3) $$R = \frac{n+2}{n}$$

Therefore $nR = n + 2$ and so $nR - n = 2$, hence

$$n(R-1) = 2$$

which yields

(3.4) $$n = \frac{2}{R-1}$$

For tetrasodium pyrophosphate, $R = 2$ and $n = 2$. For a pure compound n is exactly 2 and is not necessarily an average, but it illustrates this concept in very simple terms.

Assume that $R = 1.5$. Then $n = 4$ for tetrapolyphosphate, as a pure compound. If $R = 1.001$, then $n = 2{,}000$. Since there is no pure compound with a 2,000 chain length, it will surely represent a mixture with an average of 2,000 phosphorus atoms per chain.

This concept becomes very important when considering ways of growing chains of maximum chain lengths with an average chain length 1,000,000 or more. This means that a ratio of metal oxide (including water) to phosphorus pentoxide must be adjusted precisely in a range between 1.000002 and 1.0000000. Obviously, there is no method to accomplish this precision by merely mixing metal oxides with phosphorus pentoxide. There are methods for utilizing phase chemistry to accomplish the feat of growing very long chains, though no methods exist for growing molecules of exactly 1,000,000 average chain length, not to mention a pure compound of exactly this length.

3.1.1. Chain Breakers and Terminators

Phosphate glasses are prepared by heating a melt to temperatures of 900 °C or more, then quickly chilling the melt on metal plates. Assume that 400 g of NaH_2PO_4 is heated to 900 °C for 10 minutes and then quenched. Resulting glasses will have an *n* of approximately 100. Exact values will depend on many variables such as surface-to-volume ratios, air humidity, purity of NaH_2PO_4, rates of chilling, and many more.

There are two major reasons why average chain length is low, when one should expect from the above equations that a system with an *R* of 1.0000000 should have an *n* near 1,000,000. As noted, water is a metal oxide in phosphate systems and contributes to the numerator of *R*. It is extremely difficult to drive all water from an amorphous system, even when an amorphous system is heated to a melt. Inhomogeneity is a second reason why an *n* value can be low. There is no convenient way to homogenize a melt system sufficiently, on a molecular scale, not to have local variations in concentrations of metal oxides to phosphorus pentoxide, when melts are quenched. Note that a ring (metaphosphate) may be counted as an infinite chain, but may be of low molecular weight.

Repeat the above experiment, heat the melt to temperatures near 900 °C, but instead of pouring a melt directly on a chill plate, allow it to cool slowly to 600 °C. This melt will crystallize, usually, but not always, to a cyclical sodium trimetaphosphate, $Na_3P_3O_9$. It will be shown later that there are a variety of crystalline substances that can result from crystallization of these sodium systems, depending upon conditions. Now, if these crystals are again heated to a melt near 700 °C and chilled suddenly, *n* of the resulting glass should be somewhat greater than 150. This glass can be recrystallized and reheated to a melt and quenched again. Each recrystallization and melting will increase the chain length until a chain length slightly under 400 is obtained as a limit. Because of the importance of this concept it will be revisited from a slightly different viewpoint.

3.1.2. Heteropoly Anions

It is a simple process to form a phosphate glass. Phosphates are called glass formers. Phosphate melts are complex mixtures of reorganizing molecules. Glasses can be formed from many of these melts when they are chilled quickly. The amorphous solids are mixtures of molecules that happened to exist as they were frozen in time. No two mixtures of phosphate glasses are ever exactly the same, but their average properties are very similar.

The simplicity of forming phosphate glasses is one reason that many more articles have been published about these amorphous systems than are published about crystalline systems. The problems become difficult when attempting to study the properties of systems that are believed to contain POX bonds, where X may be S, Si, As, Sb, Mo, B, Bi, or many other oxides. Single-phase heteropoly systems form easily. Almost everything will dissolve in a phosphate glass. But the fact that an amorphous system is a homogeneous glass does not mean that the phosphates and the second substance have reacted molecularly. It is much more likely that the second system is dispersed in the phosphate glass. The chances of forming POX linkages in glasses are much greater if it is known that the linkages have been identified in crystalline compounds.

Studies of crystalline systems are definitive, because a thermodynamically defined condition may be specified and analyses by X-ray and similar techniques leave little reason to doubt the results. In other amorphous systems some properties of the systems are so dramatically different from the unreacted materials that there is no room for doubt. Such a system is the amorphous system composed of phosphates and silicates.

3.1.3. Increasing Chain Lengths of Phosphate Glasses

As mentioned above, if a system is crystallized and melted several times, average chain lengths of glass will continue to grow each time until a chain length near 400 is obtained. This is not an absolute limit, but is about as large as chains can be grown by this technique. Even if a crystalline sodium Kurrol's salt that is known to have a chain length of 5,000 or more is melted and chilled to a glass, the resulting glass will seldom have n greater than 400. The highest molecular weight sodium polyphosphate ever observed by the author was prepared by Dr. Marvin M. Crutchfield. He treated long crystals of $[Ca(PO_3)_2]_n$ fibers with sodium EDTA solutions. The chain length was great enough to cause the amorphous sodium salt to be difficult to dissolve in water. The exact chain length was never measured and provides an intriguing series of experiments yet to be performed.

Crystallization and recrystallization drives impurities out of a melt to surfaces of crystals. Each time a system is crystallized, more and more water is lost. As water is lost, bulk compositions more nearly approach an R of unity. All metal oxides in excess of unity contribute as chain terminators and breakers.

In discussing functions of polyphosphates in living systems very long chain lengths of Kurrol's salts are mentioned. Of particular interest is potassium Kurrol's salt. In discussing sodium phosphate systems, it was noted that as melts cooled, a variety of compounds ranging from sodium trimetaphosphate, a three phosphorus ring, to long-chain polyphosphates such as Maddrell's salt or sodium Kurrol's salt and others could crystallize, depending upon a melt's history. X-ray patterns are usually used to differentiate these long-chain crystalline phosphates. Some have only slight differences in molecular structure and crystal form, but will exhibit entirely different X-ray patterns.

Unlike the sodium systems, melt history has less influence on potassium Kurrol's salt systems mentioned above. Potassium trimetaphosphate is not a phase diagram entity. Potassium phosphate melts with *R* values of unity are very difficult to quench to glasses, because they crystallize very rapidly. If potassium phosphate melts are cooled slowly, long-chain polyphosphates, $[KPO_3]_n$, are the only crystals to form. No other crystals are thermodynamically stable in this portion of their phase diagram and Kurrol's salts dominate a large area of this diagram.

Potassium polyphosphates with *R* values near unity can have *n* values of 5,000 to 10,000. In this instance crystals themselves are acting as templates to grow long-chain molecules. It should be noted that one water molecule in 10,000 phosphate groups is required to convert one 10,000 chain length molecule into two molecules with an average of 5,000 phosphorus atoms per molecule. Usually, two molecules of 5,000 phosphorus atoms each does not result. Instead, an orthophosphate and a molecule containing 9,999 phosphorus atoms would result from hydrolytic degradation. Nevertheless, the resulting average chain length is 5,000. In a melt this is not as likely to happen and molecular fragment sizes would be more likely to follow laws of chance, because making and breaking of linkages happens continuously with or without chain breakers at melt temperatures.

3.1.4. Influencing Chain Lengths of Crystalline Polyphosphates

Equation (3.3) predicts average chain lengths of phosphate glasses reasonably well for short chain length batches. The equation is about as representative of these systems, as these systems are representative of the concept of average chain length. No two batches of glass with an average chain length of 100 are equal, since polyphosphate glasses are usually mixtures of poly- and metaphosphates with random structures and sizes. In both amorphous and crystalline systems an M_2O–P_2O_5 ratio *of a system* is a dominant controlling factor.

For any single molecule–ion a ratio, *R*, precisely dictates chain length. As chain lengths of polyphosphates increase other factors play a greater and greater role, however an M_2O–P_2O_5 ratio continues to dominate as a necessary but not sufficient condition to control chain lengths effectively. These other conditions include factors such as specific metals ions, M, contained in M_2O, the phase state of a phosphate (crystalline or amorphous), composition systems containing a polyphosphate, seed crystals, and thermal history.

Calcium polyphosphates, $[Ca(PO_3)_2]_n$, crystals grown in sodium ultraphosphate melts in the preparation of fibers, were grown in a manner similar to how $[KPO_3]_n$ is grown in melts of ultraphosphoric acid. These systems yield salts referred to as cross-linked Kurrol's salts. Since potassium Kurrol's salts can be easily and quickly solubilized in water containing some diverse ions, such as sodium, potassium phosphates are much more easily studied. Potassium Kurrol's salts are phosphate fibers very similar to the phosphate fibers of interest here. They have an appearance very similar to chrysotile. It is because the potassium Kurrol's salts can be solubilized with diverse sodium ions that they can be easily studied. They will be used as a pilot material for the studies of phosphate fibers, while the behavior of the $[Ca(PO_3)_2]_n$ and $[NaCa(PO_3)_3]_n$ fibers will be compared to Kurrol's salts.

Several issues need to be addressed. Three of them are:

1. A better understanding of the variables controlling chain lengths of very long crystalline polyphosphate molecule–ions is needed. How important is melt composition to crystallizing, and what is the ratio of amorphous solids to crystalline solids when a portion of a melt is difficult or impossible to crystallize?
2. There is no question whether or not cross-linking is a necessary part of a system with an ultraphosphate composition. It is a structural-composition requirement, but where does "cross-linking" reside in potassium Kurrol's salt crystallized in ultraphosphate melts? Are cross-links contained in crystalline phases and cause the formation of crystalline potassium ultraphosphates, or does a salt obey its phase diagram for a two-component $K_2O–P_2O_5$ system? (See Figure 3.1.)
3. How do seed crystals function as templates when growing long-chain polyphosphates, and what type and size of phosphate segment delivers PO_3^- from a melt phase to a crystal surface during crystallization?

Let us approach question 1 first. Assume that when preparing cross-linked Kurrol's salt, all phosphoric oxide added to crystalline $[KPO_3]_n$ enters potassium Kurrol's salt as a cross-linking agent. Assume that all phosphate crystallizes to cross-linked Kurrol's salt. If these assumptions are true, cross-linked Kurrol's salt is an ultraphosphate and not a polyphosphate because the $M_2O–P_2O_5$ must be less than unity. Moreover, there must be two different compounds present in a phase system exhibiting different X-ray patterns.

Two different potassium Kurrol's salt compounds, one cross-linked and the other linear chains, do not appear in this phase diagram, nor could a crystalline compound other than potassium Kurrol's salt be found in ultraphosphate areas of this system. Other choices are: (1) an amorphous phase is formed, or (2) some new undetected crystalline potassium ultraphosphate is formed. No other crystalline potassium ultraphosphates are known, and no new ones were discovered in this work. Both phase studies and X-ray analyses suggest that an amorphous phase is formed in *all* potassium Kurrol's salt systems crystallized in ultraphosphate phase systems.

In an ultraphosphate the number of triply connected POP linkages can be calculated from equation (3.5),

(3.5) $$N_x = (P - M)/2$$

where N_x is the number of POP linkages in excess of a metaphosphate composition,[1] P is the number of phosphorus atoms or moles, and M is the number of equivalents of ions or cationic groups associated with the phosphate; ${}^tR = M/P$ (the M_2O–P_2O_5 ratio). In a metaphosphate there is one POP linkage per phosphorus atom. It is therefore possible to calculate the number of linkages in any phosphate system if the M_2O–P_2O_5 ratio is known.

The following notation will be used to specify phases as a crystalline, glassy, or total system where ${}^tV$ = total system, ${}^gV$ = amorphous system, and ${}^cV$ = crystalline system, for $V = R, M, K, P_2O_5$, or any other variables.

3.2. Structural Considerations

All single chains in a polyphosphate system can be linked together to form double chains by adding only one PO_3 cross-linking group for each two chains. Moreover, only one PO_3 group must be added per chain to build an infinite network of connected chains. For example, assume that a Kurrol's salt with an average chain length of 5,000 is mixed with P_4O_{10} to cross-link a system. One P atom in 5001 P atoms, or 0.02% of all phosphorus in a system, must come from added P_4O_{10}. Then ${}^tR = 0.9998$. Any phosphate, PO_3, in excess of this value will tighten a cross-linked structure and should lessen its aqueous solution behavior in an idealized case. (Cross-linked Kurrol's salts produce very viscous, stringy aqueous solutions even when Kurrol's salt is dissolved to make solutions of 0.1% or less.) These facts were well recognized by Pfansteil and Iler in their classic work relating cross-linking to behavior in aqueous solutions.[2]

Polyphosphates crystallized in an ultraphosphate melt may be superlong chains of linear molecules with only "defect cross-linking."[3] Also, cross-linking may be contained in an amorphous phase, where it is very difficult to detect directly, rather than as a part of a crystal structure. Cross-linking is probably a random event and does not constitute a thermodynamically stable state, even in the absence of moisture.

Pfansteil and Iler demonstrated that as the K_2O–P_2O_5 ratio is reduced from 0.998, the influence of cross-linking in solutions of Kurrol's salt diminishes, if it is not eliminated entirely. The behavior expected for very long linear chains persists after the expected behavior for cross-linkages is gone, as judged from viscosities of aqueous solutions obtained from salts with K_2O–P_2O_5 ratios well below 0.98.

3.3. Amorphous Phase Content of Long-Chain Systems

A question of interest arises in predicting ratios of amorphous phase to crystalline phase in a Kurrol's salt system prepared in ultraphosphate regions. Since

Kurrol's salt is very slowly soluble in distilled water and ultraphosphate glasses are infinitely and quickly soluble in water, it is possible to estimate the quantity of amorphous phase contained in a well-crystallized Kurrol's salt system. Sample mixtures of crystalline and amorphous phases are first milled to a fine powder and then any soluble phase dissolved from crystalline phases with distilled water. The aqueous filtrate solution is then analyzed both for ratios and concentrations of dissolved phosphates. The soluble part of a sample should be amorphous since potassium Kurrol's salt is almost insoluble under these conditions.

3.3.1. Theoretical Considerations

It is the objective of this section to calculate quantities of amorphous solid contained in an ideal potassium Kurrol's salt system in which ${}^{t}R$, the M_2O–P_2O_5 ratio of a total system, is less than unity.

1. It is known that at least one crystalline phase, $[KPO_3]_n$, is found in systems with ratios $1 > K_2O/P_2O_5 > 0.5$. Only one crystalline phase can be detected by X-ray analyses.
2. All attempts to form crystals in ultraphosphate systems with K_2O–P_2O_5 ratios of 0.5 or less have been unsuccessful. Thermal analyses reveal an endothermic region extending from about 200 °C to the "eutectic temperature" at 450 °C. This amorphous region becomes more and more pronounced as M_2O–P_2O_5 ratios are decreased to 0.5, where the total system is amorphous.
3. No potassium *ultraphosphates* in the composition range $1 > K_2O/P_2O_5 > 0.5$ are crystalline.
4. All ultraphosphate systems considered here have a K_2O–P_2O_5 mole ratio, ${}^{t}R$, such that $1 > {}^{t}R > 0.5$.

3.3.2. Predicting Amorphous Content of Crystalline Systems

It is desired to derive equations relating crystal phase, amorphous phase, and total composition of systems studied. The derivation is simpler if it is assumed that one mole of K_2O is contained in a total system and is divided between crystalline and amorphous phases:

$$ {}^{t}R = {}^{t}K_2O/{}^{t}P_2O_5 \tag{3.6} $$

For ease of calculation consider two moles of KPO_3 as $K_2O \cdot P_2O_5$. Add z moles of P_2O_5 to two moles of KPO_3 to adjust the ratio, ${}^{t}R$:

$$ {}^{t}R = \frac{K_2O}{P_2O_5 + z} \tag{3.7} $$

One mole of K_2O is distributed between crystalline phases and amorphous phases. Therefore

(3.8) $${}^{g}K_2O + {}^{c}K_2O = 1$$

(3.9) $${}^{g}P_2O_5 + {}^{c}P_2O_5 = 1 + z$$

and

(3.10) $${}^{t}R = \frac{1}{1+z}$$

where ${}^{c}R = 1$, by definition, if KPO_3 is the only crystalline phase. Also

(3.11) $${}^{g}R = \frac{{}^{g}K_2O}{{}^{g}P_2O_5} = \frac{{}^{g}K_2O}{1+z-{}^{c}P_2O_5} = \frac{1-{}^{c}K_2O}{1+z-{}^{c}P_2O_5}$$

When it is assumed that $[KPO_3]_n$ is the only crystalline phase, then the amorphous phase must contain at least z moles of P_2O_5. Therefore the amorphous phase must contain

(3.12) $$zP_2O_5 + yP_2O_5 + yK_2O \text{ moles (gram-ions)}$$

where y equals the moles of $2KPO_3$ dissolved into the amorphous phase;

(3.13) $${}^{g}R = \frac{y}{z+y}$$

(3.14) $$y = \frac{{}^{g}R}{1-{}^{g}R}z$$

When ${}^{g}R = 0.5$, $y = z$ and the total system is amorphous. This agrees with experimental results. The phase status of a solidified melt is controlled by the K_2O–P_2O_5 ratio, ${}^{t}R$, and a sample's thermal history. Recalling that z is moles of P_2O_5 added per two moles of KPO_3, it can be shown that

(3.15) $$z = \frac{1 = {}^{t}R}{{}^{t}R}$$

and therefore

(3.16) $$y = \frac{{}^{g}R(1-{}^{t}R)}{{}^{t}R(1-{}^{t}R)}$$

When ${}^{t}R = 0.5$, ${}^{g}R = 0.5$, and $y = 1$, the conditions for a completely amorphous system are satisfied. Then, under these conditions,

(3.17) $$y = \frac{1 - {}^{t}R}{{}^{t}R} = z$$

It is possible to predict the percent amorphous phase formed in these systems, provided the ratio ${}^{g}R$ is determined. From equation (3.6)

(3.18) $$z = \frac{1 - {}^{t}R}{{}^{t}R}$$

therefore

(3.19) $$y = \frac{{}^{g}R}{1 - {}^{g}R} \times \frac{1 - {}^{t}R}{{}^{t}R}$$

The weight percent glass in the crystallized system is then

(3.20) $$\%\ \text{Amorphous} = \frac{236.14\, y + 141.94\, z}{236.14 + 141.94\, z} \times 100$$

where the numbers are formula weights of $2KPO_3$ = 236.14 and of P_2O_5 = 141.96.

3.3.3. Comparison of Theoretical Values and Analyses of Kurrol's Salts

To obtain chemical analyses and phase separations samples of Kurrol's salts weighing 20 g, with K_2O–P_2O_5 ratios ranging from 1.00 to 0.60, were stirred in 480 g of distilled water for 20 hours at room temperature. Three methods were employed (I, II, III) to determine how much of the sample dissolved. Aqueous phases were filtered from crystalline Kurrol's salt and K_2O–P_2O_5 ratios and concentrations of dissolved salt were determined.

Method I: Filtered residue solids were air dried for 2 days and then weighed. Percent solubles were calculated from differences.

Method II: A precisely weighted sample of filtered solution was mixed with a pre-ignited crucible containing ZnO and dried at 110 °C for 1 hour, followed by heating at 550 °C for 30 minutes. Cooled, desiccated samples were again weighed at room temperature.

Method III: Samples of filtered liquids were analyzed by inductive coupling plasmas and atomic absorption. These results are presented in Table 3.1.

Table 3.1 compares the measured values with predicted values for the quantity of amorphous phase formed. It is noteworthy that even at a ratio as large as 0.98 a relatively large percent of amorphous phase (7.96%) should be formed in a potassium Kurrol's salt system. Measured analytical results support these predictions.

Calculated results are in remarkably good agreement with analyzed values. Solubles in sample of $[KPO_3]_n$ from systems with a ratio of unity suggest that samples were not exactly unity or that they were not homogeneous. In other cases

Table 3.1. A Comparison of the Quantity of Amorphous Phase by Analysis and by Calculation

		Measured % amorphous			
^{t}R Total system	^{g}R Amorphous phase	I	II	III	Calculated % amorphous
1.00	1.00[a]	2.20	2.55	1.17	0.00
0.98	0.77	7.50	8.92	7.19	7.96
0.94	0.77	29.65	29.17	29.59	24.27
0.75	0.66	72.90	75.05	73.79	70.60
0.65	0.61	88.22	88.77	87.86	88.08
0.60	0.59	96.67	96.04	95.93	97.10
0.50	0.50		—		100.00

[a]Analyzed value.

calculated values are probably better than analytically determined values when ratios of solubles are known precisely.

When sufficiently large quantities of P_2O_5, z, are added to react with all crystalline $K_2O{\cdot}P_2O_5$ in a system, the reaction between potassium Kurrol's salt and P_2O_5 may be written as

$$P_2O_5 + 2[KPO_3] = K_2O{\cdot}2P_2O_5 \text{ (glass)} \tag{3.21}$$

Each gram of P_4O_{10} added to 1.66 grams or more potassium Kurrol's salt forms 2.66 grams of amorphous phase when the system is heated to a melt and allowed to cool to glass. Equation (3.21) is unique for preparations of a system with a ratio of 0.50.

3.4. Phase Diagram and Cross-Linkages

A discussion of phase diagrams for pertinent phosphate systems is given in Chapter 5. Whether or not a presentation of the K_2O–P_2O_5 phase diagram should be delayed until Chapter 5 was considered. If difficulties are encountered, hopefully, referring to Chapter 5 will be helpful.

A nonequilibrium extension to the K_2O–P_2O_5 phase diagram[4] requires additional work to understand this system. (See Figure 3.1.) Our extension cannot qualify as a *bona fide* phase diagram, because there is no claim that this system is at equilibrium. It is surely not at equilibrium. Some behavior of this system can be gleaned from the diagram and it should represent some behavior reasonably well. Our attention will be focused upon the left side of this diagram until short chain length Kurrol's salts are discussed below.

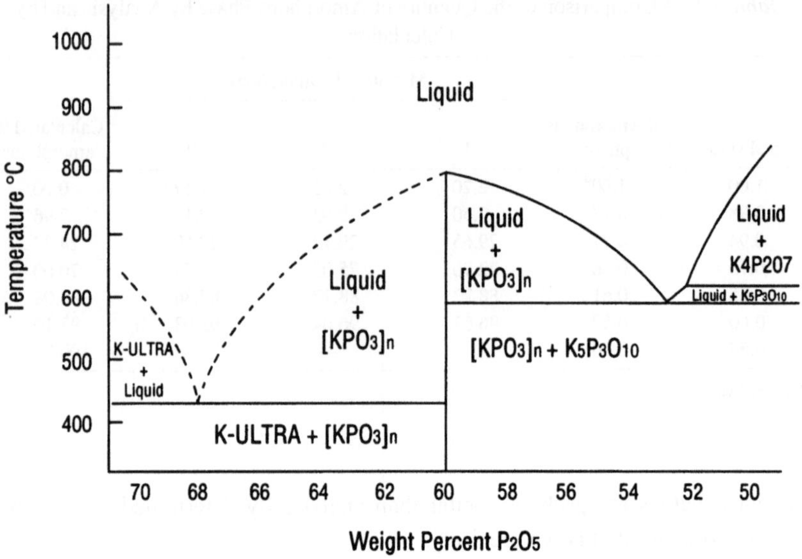

Figure 3.1. Potassium phosphate phase diagram plus the nonequilibrium extension.

A eutectic behavior in the P_2O_5-rich portion of this diagram is probably a melt solubility temperature with $[KPO_3]_n$ being the only potassium containing crystalline phase. Pure P_2O_5 can be both crystalline or amorphous and is far to the left of the left-hand axis. The hexagonal form of P_2O_5 has a sublimation temperature of about 360 °C and the "liquidus" is probably an artifact of the rate of heating and probably represents sublimation or boiling. Extrapolation of this thermal data cannot represent a melting point of P_2O_5. All samples of Kurrol's salt subjected to thermal analysis (heating curves only) exhibited an endothermic reaction in the temperature range from 200 °C to 450 °C. This reaction is believed to be thermal mobilizing of an amorphous phase and solubilizing of crystalline $[KPO_3]_n$ at 450 °C, where there is a strong endothermic reaction.

No potassium phosphate crystals of any kind can be formed in an anhydrous system with a K_2O–P_2O_5 ratio less than 0.5. It was shown above that a 0.5 ratio is the minimum ratio at which $[KPO_3]_n$ crystals can form. The "eutectic" behavior is the dissolution of $[KPO_3]_n$ crystals in a melt with a ratio near the "eutectic" composition when the system contains less than 68% P_2O_5. When all the amorphous phase with a P_2O_5 concentration greater than the "eutectic" composition is depleted, temperatures must be increased to dissolve the remaining $[KPO_3]_n$ crystals in a melt that is richer in K_2O than the "eutectic" composition. This is an explanation of the behavior of this system as determined, but leaves unanswered questions as to exactly what is happening and whether *crystalline* Kurrol's salts contain cross-linkages. In turn, this is related to the question whether or not $[Ca(PO_3)_2]_n$ and $[NaCa(PO_3)_3]_n$ crystals contain cross-linkages of the phosphate chains in the crystals.

It is concluded that cross-linking of Kurrol's salts is a random transient event that occurs because the system is not at equilibrium. Allowed sufficient time, these systems will become mixtures of linear $[KPO_3]_n$ molecules and an amorphous phase. A sample that was held at temperatures just below the melting point for five days lost much of the exaggerated aqueous solution properties usually associated with cross-linked Kurrol's salts.

Although the above approach is easy and straightforward, care must be exercised when applying it to more complex systems. Growth of calcium polyphosphates in sodium ultraphosphate melts is a three-component system rather than two, even when all water is driven from the system. It is recommended that the simpler method of assuming all sodium and no calcium reported to an amorphous phase (as was done in Chapter 6, Section 6.5) be used to estimate ratios of amorphous to crystalline phases or portions of a melt that is expected to be water soluble when the melt is fully crystallized. These estimates will not be precise, but are useful in estimating potential yields.

3.5. Short-Chain Kurrol's Salts

Attempts have been made to grow short chain length Kurrol's salts from melts with M_2O–P_2O_5 ratios between 1.04 ($\bar{n} = 50$) and 1.10 ($\bar{n} = 20$). If a slight excess of Na_2O or K_2O behaved on the basic side of metaphosphate compositions, as P_2O_5 is reported to behave on acid sides of metaphosphate compositions, it should be possible to grow short-chain Kurrol's salts as well as cross-linked Kurrol's salts. The required quantities of $Na_5P_3O_{10}$ and $K_5P_3O_{10}$ always crystallized from melts as required by the phase diagram, while Kurrol's salt was very similar, if not identical, with salt obtained from melts with M_2O–$P_2O_5 = 1.0$.

The phase diagram to the right of metaphosphate compositions is an equilibrium diagram and is much more trustworthy than the extension presented above. Until now no one has been capable of growing a Kurrol's salt, sodium or potassium, with an average chain length of 50 to 100 if the tripolyphosphate had been removed from the Kurrol's salt. Naturally, if an M_2O–P_2O_5 ratio had been adjusted to yield an average of 50 the mixture of tripolyphosphate and Kurrol's salt would have averaged 50, but only because tripolyphosphate would have acted to yield a lower average. Kurrol's salt fractions grown from basic melts was practically unchanged from Kurrol's salts grown from acidic melts.

3.6. Triply Linked Phosphate Groups

If crystalline inorganic phosphates grown in ultraphosphate media contain triply linked phosphate groups, they are either isopolyphosphates or ultraphosphates. An example is a PO_3 group bonded through an oxygen linkage to three other PO_3 groups. Three types of triply linked phosphate groups are possible.

Type I: Cross-linking

```
  O O O O O O O
 –O POPOPOPOPOPO–        (charges are ignored)
  O O O O O O O
       OPO
  O O O O O O O
 –O POPOPOPOPOPO–
  O O O O O O O
```

where there are two triply linked PO_3 groups, but there are two other types of bonding for branching and cross-linking. In the structure shown above there is a single PO_3 in the bridge. There is no reason why there could not be several PO_3 groups between the two longer chains. It is also possible that there could be no PO_3 between the chains as in the following structure.

Type II: Cross-linking

```
  O O O O O O O
 –O POPOPOPOPOPO–        (charges are ignored)
  O O  |  O O O
       O
  O O  |  O O O
 –O POPOPOPOPOPO–
  O O O O O O O
```

where there are but two triply linked PO_3 groups.

There is then branched isopolyphosphate that contains a single triply-linked PO_3 group as below.

Type III: Branching

```
  O O O O O O O
 –O POPOPOPOPOPO–        (charges are ignored)
  O O O O O O O
       OPO
        O
       OPO
        O
       OPO
        |
```

In an attempt to obtain direct evidence for the presence of cross-linking in potassium polyphosphates crystallized from ultraphosphates melts, two tools were chosen. Cross-linking should in theory be observed by either phase chemistry or X-rays. Also, the existence of end and middle group phosphorus is well established in NMR studies of inorganic phosphates while triply linked phosphate groups have been reported in organic phosphate systems. Cross-linking in polyphosphates should be directly observable by tunneling microscopy.

3.7. Seed Crystals as Templates

If polyphosphates of chain lengths greater than about 350 are to be obtained, a crystallization process is required in which a seed crystal is behaving as a template for chain growth. Templates aid crystallization to proceed without chains of molten polyphosphate "sequestering" metal ions to terminate their growth or thermal activity rupturing molecules.

To obtain high tensile strengths of calcium polyphosphate fiber crystals, it is necessary to grow polyphosphate molecule–ions as long as possible. When a phosphate group from a high-temperature melt system moves to a high-temperature crystal lattice, a phosphate group has transgressed a phase boundary. In a phosphate melt system molecules are reorganizing chaotically. In a crystal lattice environment a system is highly organized. A phosphate group or groups must not only fall to a lattice of lower chemical potential, but must also be accompanied by the correct number of cationic groups to exactly balance the negative charge on the phosphate. This must be done without the polyphosphate sequestering enough metal oxide to act as either chain terminators or breakers.

Longer-chain polyphosphates grow to greater lengths in systems deficient in metal oxides containing an excess of P_2O_5. Under these conditions metal oxides will be attracted to a more acidic melt phase and this melt phase will exercise an element of control over a crystallizing polyphosphate. It is expected that longer-chain polyphosphate molecules should grow in crystals where the melts have the ratio, $M_2O/P_2O_5 < 1$, but not so low as to move into a region of the phase diagram dominated by other compounds.

Condensed phosphate melts are composed of many phosphate species that are undergoing constant reorganization, as judged by analyses of glasses formed by quenching these melts. When crystals form in high-temperature systems, segments of phosphates that transfer from melt phases to crystalline phases are limited to four models.

1. Crystals may grow one PO_3 group at a time under which conditions a molecule–ion of 10,000 phosphate groups would require the formation of 9,999 P–O–P linkages on a surface of a crystal to create a single molecule–ion.
2. PO_3 groups could be transported to crystal surfaces in clusters of three, as stable triphosphate chains or rings.
3. PO_3 groups could be transported as short to intermediate length, phosphate chains or rings, as indicated by some physical data.

4. Crystals could grow from very large molecule–ions, 10,000 PO_3 long, in melt phases and crystals could grow from single segments.

To consider each model it is unlikely that single monometaphosphate or orthophosphate groups are formed and then transported from a melt to a crystalline phase. Glasses formed from condensed phosphate melts contain no detectable orthophosphate when analyzed soon after they are prepared.[5]

There are several reasons why segments as short as three phosphate groups could be involved in a transfer of PO_3 groups from a melt phase to a crystalline phase of a long-chain polyphosphate. As noted above, when quenched, potassium phosphate melts yield rings that are not found in the crystallized systems. Tripolyphosphate which results from ring opening is too short to form knots. It is well known that polyphosphates reorganize to cyclic trimetaphosphates in aqueous solutions, but this is yet another mechanism.[6]

Figure 3.2 is an artist's conception of phosphate anions making and breaking in a melt phase. In Figure 3.3 an anion fragment is attaching itself to a surface of a crystal. This crystal is at a temperature just below the melting temperature of the crystal, and the anion that is attaching itself must give up its heat of crystallization to remain attached. Otherwise, an anion would merely bounce back into a melt region. It is possible that this energy is sufficient to rupture an anion. A fragment

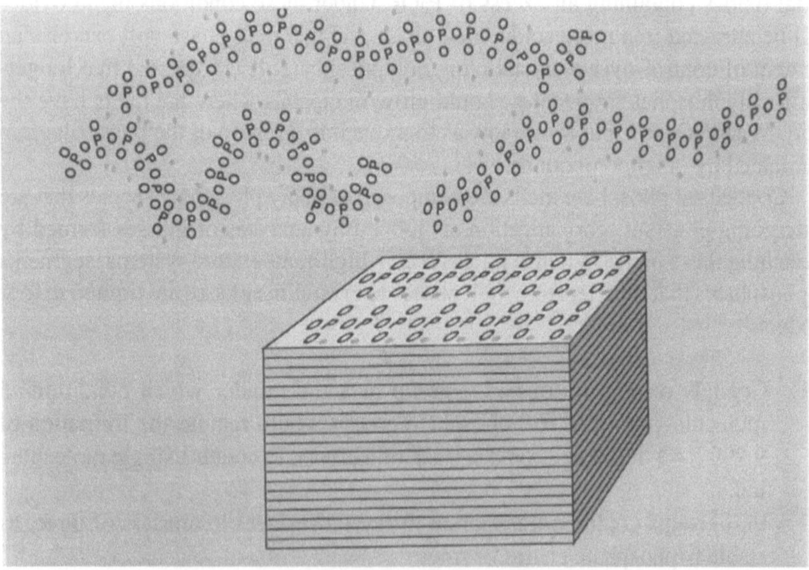

Figure 3.2. Molecules making and breaking POP linkages in a phosphate melt.

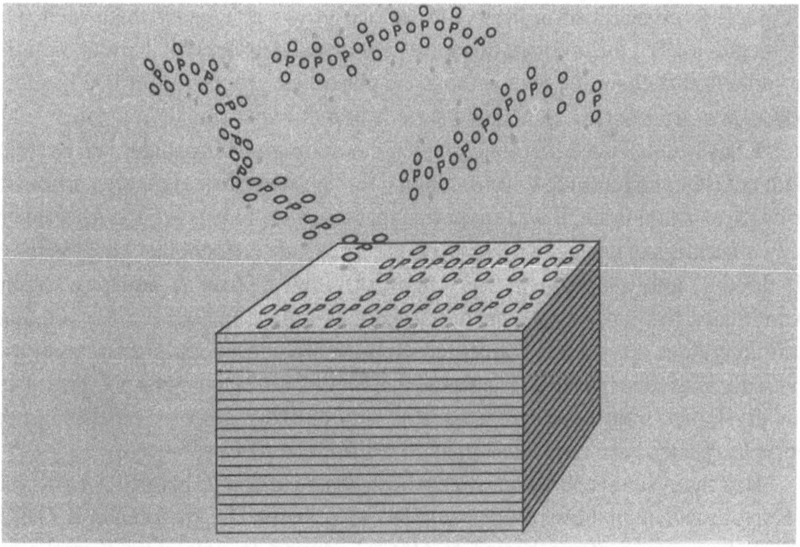

Figure 3.3. Molecule crystallizing on a polyphosphate crystal's surface.

remains on the crystal while yet another portion returns to the melt phase carrying any excess energy with it.

It is well known in crystallizations, particularly in solutions, that crystallization is dynamic and crystals are continuously dissolving and redepositing. Small crystals will disappear while larger crystals will grow even larger at the expense of smaller ones. This is another example of influences of surface-to-volume ratios often mentioned in discussions on behaviors of crystals and melts.

Surface tension and density work suggest that transfer units may range from dimers to as many as twelve phosphate groups held together through POP linkages. The most likely conditions are that the phosphate is transported as segments of six to eight phosphate groups. It was demonstrated by viscosity measurements of polyphosphate melts that the flow segment is between six and eight phosphate groups in length.[7] Segments of this length are short enough and rigid enough to be free of entanglement with other chains, and should add to a chain growing on the surface of a crystal without encountering large stearic barriers.

Convincing experiments left little doubt that most amorphous polyphosphate glasses in sodium systems exhibit some branching.[8] It is unlikely that long-chain branched structures could fit into a crystal lattice without first rupturing to destroy a branching point. It is well established that melts with $M_2O–P_2O_5 \geq 1$ resist branching.[9]

During crystallization, a crystal surface grows to contact fresh melt while molecule–ions of the melt are transported to crystal surfaces by thermal motion. It is unlikely that chains of several thousand phosphate groups are formed in a liquid phase, then transported, intact, to a surface of a growing crystal template.

It has been claimed that long-chain ammonium polyphosphates can be treated with potassium hydroxide to form long-chain polyphosphates that crystallize from solution as a dihydrate. It was suggested that the phosphate was a Kurrol's salt, but only after this salt had been heated to 600 °C.[10] There is doubt that long chains can fit directly into a crystal lattice when crystallized from an aqueous solution. Admittedly, the chaotic reorganization that takes place in high-temperature melts definitely does not occur in aqueous solutions at atmospheric pressure, but more evidence is needed before it is accepted that long-chain alkali metal polyphosphates can crystallize from aqueous media. It is very unlikely, since no other long-chain polyphosphates have ever been crystallized from aqueous solutions.

Had the system reported above not been heated to 600 °C before it was declared a Kurrol's salt, then these results could be more acceptable. As noted in K_2O–P_2O_5 phase diagrams, systems heated to 600 °C convert to potassium Kurrol's salts irrespective of the degree of polymerization of an ammonium feed stock. Exactly the same results would have been obtained had the starting raw materials been mono- or diammonium orthophosphates.

Long-chain polyphosphates grown as crystals form a narrower distribution of chain lengths than those of amorphous systems when measured by intrinsic viscosity or gel chromatography.[11] A narrow distribution would be required because molecules must fit an orderly lattice, but it has been pointed out that a single long-chain polyphosphate may reside in 10,000 or more unit cells.[9] End groups occur so seldom, is it possible that a crystal can accommodate small numbers of dislocations without changing an X-ray pattern of a phosphate?

3.8. Crystals of Potassium Kurrol's Salt

There is little reorganization in a crystal of potassium Kurrol's salt even at temperatures near melt temperatures, because there are no phase transitions in potassium Kurrol's salt crystals. Random cross-linkages may form and disappear as equilibrium is reached. In sodium systems, crystals of Kurrol's salt do suffer phase transitions. At a phase transition, chaos of melts is recaptured over an interval of transition.

X-ray patterns of crystalline long-chain potassium Kurrol's salts are highly reproducible, although phosphates can be crystallized from a variety of melt compositions. An outstanding example of this behavior is growth of potassium Kurrol's salt in more basic polyphosphate phase regions and potassium Kurrol's salt grown in acidic ultraphosphate phase regions.[12] Solution properties of long chain length phosphates are very different depending on compositions of melts from which crystals were grown, and yet X-ray patterns are precisely the same.

3.9. Aluminum Polyphosphates

Aluminum polyphosphates are currently being considered as amorphous phosphate fibers. These glasses have some interesting high-temperature properties and are worthy of consideration as phosphate fibers. The aluminum orthophosphate monomers can be prepared as water-soluble solutions that are cured to amorphous polyphosphate solids at temperatures near 500 °C and are more refractory than most condensed phosphates. After curing, these glass fibers can be heated to temperatures well above 1000 °C without decomposing.

The aluminum orthophosphate solutions are typical of yet another type of bonding that can be regarded as ionic cross-linking. When multivalent cations such as ferric or aluminum ions are components of a phosphate melt or glass, it is possible to consider phosphates linked by an ionic $PO^- \ Al^{+3} \ ^-OP$ type bond as well as a covalent POP linkage. Aluminum orthophosphates are sometimes called ionic polymers.[9] And it was a crystalline $Al(P_3O_9)_3$, aluminum trimetaphosphate, that Linus Pauling used for his classic single-crystal X-ray studies.[13] These systems may be prepared as either crystals or glasses, depending on conditions and compositions. If systems are to remain vitreous for long periods, special care should be taken to prepare polyphosphate systems and not metaphosphates or ultraphosphates. Ring aluminum metaphosphates are known as crystalline phosphates, while the ultraphosphates are inclined to be hydrolytically unstable.

Some interesting chemistry emerges when aluminum or other trivalent ions are incorporated as cations in a phosphate system. An aluminum ion, Al^{+3} represents $M = 3$. This introduces complexities in considering what the structure of an aluminum phosphate is when an Al_2O_3–P_2O_5 ratio is equal to one, one-half or one-third, as examples. Consider how one can make a polyphosphate glass that will remain amorphous, insoluble in water, and not be hygroscopic. There are many systems approaching these requirements, but if an optimum in properties is to be obtained, Al_2O_3–P_2O_5 ratios must be greater than one-third. This means that the M_2O–P_2O_5 ratio must be greater than unity, but less than two to avoid the crystallization of pyrophosphate or tripolyphosphates.

An aluminum phosphate ionic polymer is nothing more than a complex of aluminum ions and orthophosphate ions (Figure 3.4). At a ratio near one Al^{+3} to three PO_4^{-3} derived from Al_2O_3 and H_3PO_4, the pH of an aqueous solution is much lower than is expected from experience with alkali metal phosphates, where a pH of a dilute solution is near 4.5. A similar aluminum orthophosphate solution with an Al_2O_3–P_2O_5 ratio near 1/3 has a pH near 2. An aluminum ion effectively drives H^+ ions from the $H_2PO_4^-$ ions by complex formation while remaining water-soluble. This is very similar to the action of H_3PO_4 used in the old Zimmerman–Rhinehart

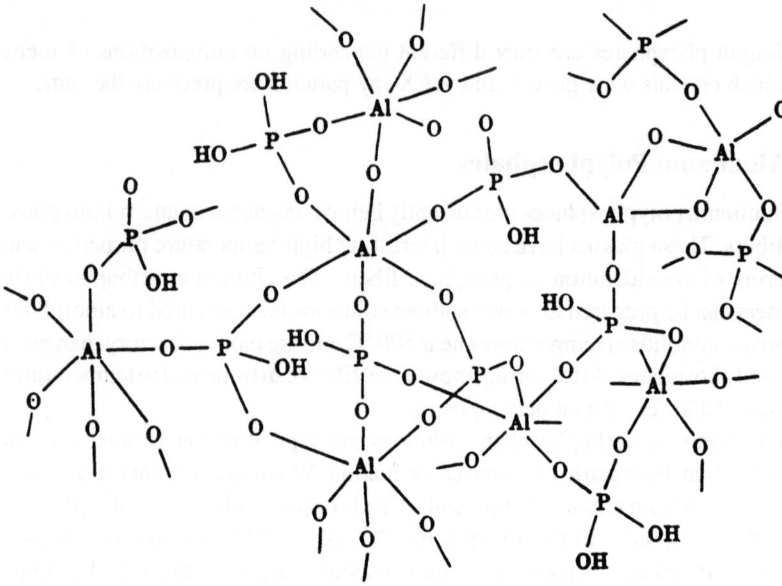

Figure 3.4. Ionic orthophosphate polymer as conceived by J. R. Van Wazer.

analysis for ferrous iron when potassium permanganate was used as an oxidizing reagent. Phosphoric acid complexes with ferric ions making them colorless, thus allowing an end point of permanganate to be seen easily.

Crudely, an equation can be written that ignores the stability complex for the ($Al^{+3} \cdot PO_4^{-3}$) complex:

(3.22) $$Al^{+3} + H^{+1} + H_2PO_4^{-1} \Rightarrow 3H^{+1} + (Al^{+3} \cdot PO_4^{-3})$$

The chemistry is surely not this simple and equilibrium constants have been determined for the various ions that might exist. At best the chemistry is a bit fuzzy. It is not our purpose to handle the details, but merely to suggest why the solutions are much more acidic that those encountered in most soluble salt solutions with similar degrees of neutralization.

Another way to view the behavior may be related to another old analytical technique of pH titrations to determine the concentration of orthophosphate in a solution by causing a precipitation of Ag_3PO_4 by the reaction by the Gerber Miles titration.

(3.23) $$3Ag^{+1} + HPO_4^{-2} \Rightarrow \underset{\text{(solid)}}{Ag_3PO_4} + H^{+1}$$

Here the released H^{+1} ion is titratable. The acid, HPO_4^{-2}, is too weak to titrate. This is but another way to release free hydrogen ions in a solution with neither hydrolysis nor soluble complex formation.

Greger considered these ionic polymer systems to be colloids, and indeed they do possess behavior similar to hydrophylic colloids.[14] Because of the complexity of the strongly acidic, ionic-covalent poly nature of the system, no definitive experiments have been made to determine whether or not the solutions are true solutions or colloidal dispersions. Perhaps light scattering would be one of the simpler approaches, but I am unaware of this technique being applied to the system.

If the solution of ionic polymer is heated at temperatures not to exceed about 100 °C, water of solution is driven from the system producing a solid, amorphous, ionic polymer. If this polymer is subjected to atmospheres of high vapor tensions, the polymer will absorb water to deliquescence and the solution will ultimately be diluted, because the vapor pressure of this solution is equal to the vapor tension of its environment. If, on the other hand, an ionic polymer is heated to temperatures near 400 °C a covalent-ionic polymer is formed of the type seen in Figure 3.5. In this case segments of polyphosphates are held together by covalent linkages, while the assemblies are held by ionic attractions between aluminum ions and polyphosphate moieties.

The amorphous solids obtained by heating ionic polymers of aluminum phosphate are not hygroscopic to the degree that purely orthophosphate polymers are. They show little tendency to dissolve or absorb more than surface water from high humidity atmospheres. This is a result of the "lattice" of this solid, that is an infinite three-dimensional network of polyphosphates uniting at triple intersections of aluminum ion terminals.

To obtain a polyphosphate lattice that has a minimum chance of crystallizing to aluminum trimetaphosphate, AlP_3O_9, it is better to lock chains in a three-dimensional network. Assume a composition is required in which, on average, 100 PO_3^- groups are held together by covalent POP linkages and terminated by an Al^{+3} in which three chains terminate at a single point. All remaining Al^{+3} ions are distributed through the mass to neutralize negative charges along chains in a type of ionic cross-linking. Here, on average, each phosphate chain will have 99 PO_3^- and one PO_4^{-2} groups.

One may use equation (3.4) to calculate required ratios of aluminum oxide to phosphorus pentoxide. If n is to equal 100, then R must equal 1.02. This will insure that all required properties will be met when a system is heated to an extent that all water is driven from the system. If these conditions are to be obtained, then all end group water must be driven from a polyphosphate system.

Consider what would happen if considerably more Al_2O_3 were added to a system than is required by the stipulated conditions. If as much as one Al^{+3} were added per PO_4^{-3}, then the M_2O–P_2O_5 ratio would be three and only an orthophosphate $AlPO_4$ could be obtained. Conversely, if ratios of Al to P were 1:2, as is sometimes written in patent literature, an M_2O–P_2O_5 ratio is 3:2 or 1.5 and n is four.

If a mole ratio of aluminum oxide to phosphorus pentoxide is slightly less that 1/3, phosphate structure changes dramatically as noted above with cross-linked

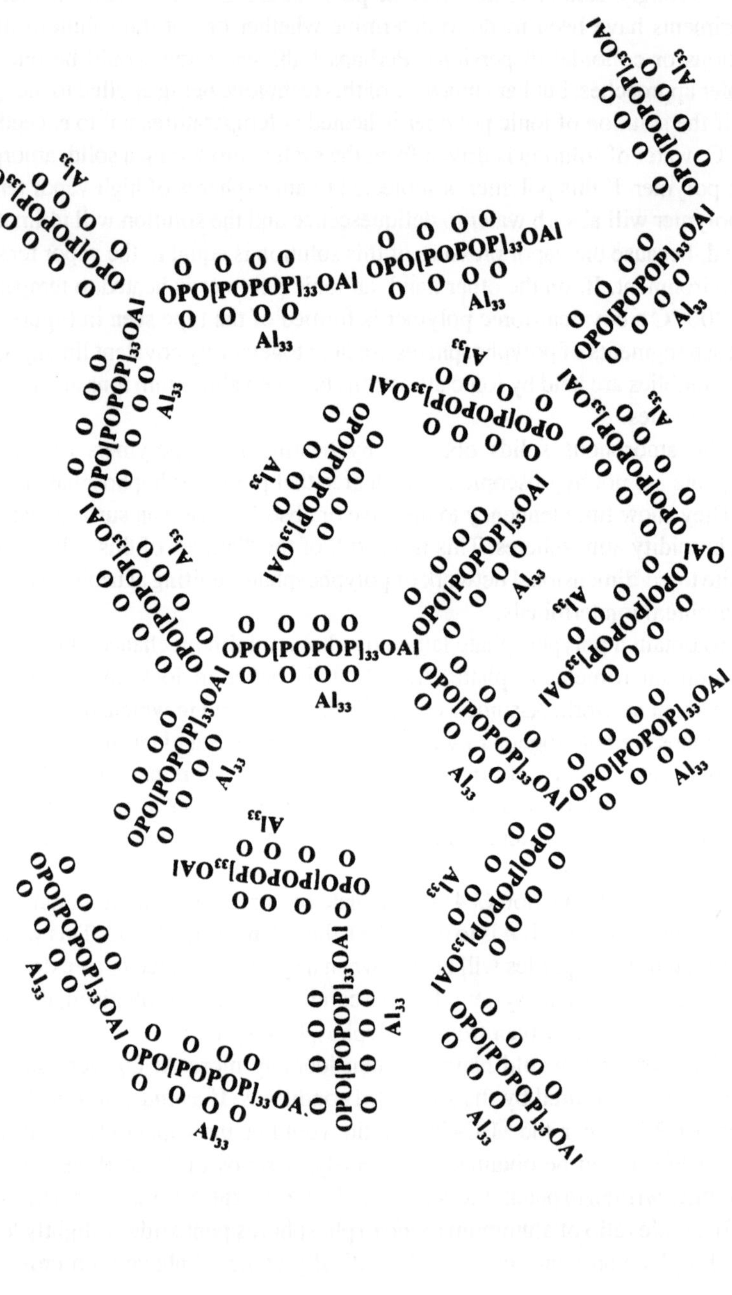

Figure 3.5. Covalent polymer network with average chain length near 100 after heating to drive out water of constitution.

potassium Kurrol's salt and cross-linkages are built into an amorphous "lattice." It is not a lattice in the sense of a crystal lattice, but some molecular structure is required. A move into ultraphosphate regions is destined to cause materials to become highly hygroscopic and ultimately soluble. When a system is amorphous almost no triply linked phosphate groups can be allowed, if initially required conditions are to be achieved.

The overall reactions may be considered as below. In equation (3.24) the simple ionic polymer is formed.

$$Al_2O_3 + 6H_3PO_4 \Rightarrow 2Al^{+3} + 6H_2PO_4^{-1} + 3H_2O \quad (3.24)$$

$$Al^{+3} + 3H_2PO_4^{-1} \Leftrightarrow (Al^{+3} \cdot 3HPO_4^{-2}) + 3H^+ \quad (3.25)$$

With heat below,

$$n(Al^{+3} \cdot 3HPO_4^{-2}) + 3nH^+ \Rightarrow [Al(PO_3)_3]_n + 3nH_2O \quad (3.26)$$

Let us consider the general structure of the product of equation (3.26) as units of three PO_3 units. There will be 33 tri-units in the average moiety and they will combine as follows:

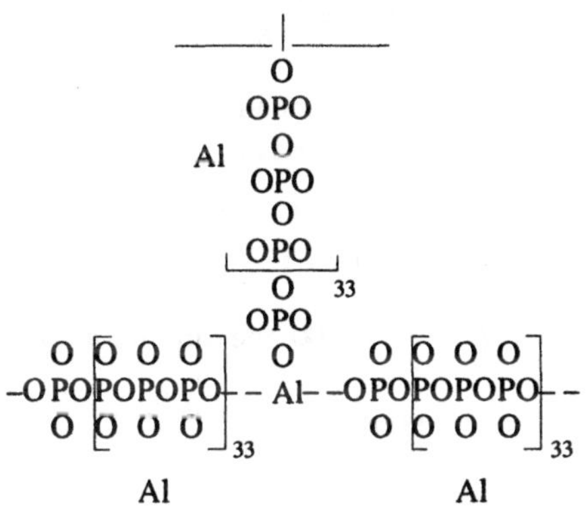

These units combine to form a three-dimensional network of the type shown in Figure 3.5. This network becomes almost insoluble and may be used to make aluminum phosphate fibers with very high pour temperatures. Some limited work suggests that temperatures over 1500 °C may be reached with some fibers.

$AlPO_4$ Crystallization

$AlPO_4$ has properties remarkably similar to SiO_2, as mentioned earlier.[15] An interesting exercise results from exploring the change in chain length of the segments of POP linkages if $AlPO_4$ is induced to crystallize from an aluminum polyphosphate melt. To illustrate how the formation of an enamel of $AlPO_4$ and an amorphous matrix can cause the formation of POP linkages in the amorphous matrix as a result of lowering the M_2O–P_2O_5 ratio of the matrix, consider the following example. It is, of course, again recognized that ${}^{\prime}R$ remains constant and the crystallization of $AlPO_4$ merely removes relatively more Al_2O_3 than P_2O_5 from the amorphous phase.

Example 1. Assume we choose to form an average chain length near 50 or above when the aluminum phosphate glass is cured to cause crystallization of $AlPO_4$. From equation (3.1)

$$M/P = \frac{n+1}{n}$$

Assume also that we start with a composition that would have an average chain length, n, of 16, just to keep things simple. Then,

$$M/P = \frac{16+2}{16} = \frac{18}{16}$$

But Al = 3M; therefore

$$\text{Al}/P = \frac{18/3}{16} = \frac{6}{16}$$

If one Al reacts with one PO_4 to form crystalline insoluble $AlPO_4$, one Al and one PO_4 are removed from the amorphous phase and the ratio of the amorphous part becomes

$$\text{Al}/P = \frac{6-1}{16-1} = \frac{5}{15} = 1/3$$

However Al = 3M, and M/P = 3/3 = 1.00, the ratio for an infinite chain.

$$M/P = 1.00 = \frac{n+2}{n}$$

$n = \infty$, which represents very long chains.

Example 2. Start with the same ratio for n = 16, but assume only 2/3 as many aluminum ions precipitate on heating the phosphate solution. Multiply both the numerator and denominator of 6/16 by 3 to obtain 18/48, which is the same thing.

This time assume that 2 of 18 Al ions react to form $AlPO_4$. Then

$$\text{Al}/P = 16/46 \qquad \text{and} \qquad M/P = 48/46$$

$$\frac{48}{46} = \frac{n+2}{n}$$

$$48n = 46n + 92$$

$$2n = 92 \qquad \text{or } n = 46$$

which is near 50.

One could adjust the starting ratios by trial and error until it was learned which starting ratio would yield the proper amount of precipitated $AlPO_4$ to give an amorphous phase with the desired ratio of Al_2O_3 to P_2O_5. It should be possible to repeat the above Kurrol's salt type derivation to predict the ratio of glass to crystalline $AlPO_4$. It is assumed that $AlPO_4$ is the only crystalline phosphate in the metaphosphate to orthophosphate phase diagram, exclusive of aluminum trimetaphosphate. This may or may not be true, but it should not influence these calculations.

Start with one mole of crystalline $AlPO_4$. Add P_2O_5 to form a glass. It can be assumed that all added P_2O_5 reports to the amorphous phase as ${}^{t}R$ is lowered from 3.00 toward 1.00. Al/P is lowered from 1.00 toward 0.333, or less.

Reasoning from equations (3.6) and (3.7), we note that the Al equivalency of 3 must be considered,

$${}^{t}R = \frac{3 \times \text{Al}}{P + z}$$

where z is the number of moles of P added as P_2O_5.

$${}^{g}\text{Al} + {}^{c}\text{Al} = 1.00$$

$${}^{g}\text{P} + {}^{c}\text{P} = 1.00 + z$$

$${}^{t}R = \frac{3}{1 + z}$$

$${}^{c}R = 3$$

From equation (3.11),

$${}^{c}R = \frac{3{}^{g}\text{Al}}{{}^{g}\text{P}} = \frac{3 - 3{}^{c}\text{Al}}{1 + z - {}^{c}\text{P}}$$

If y equals the moles of $AlPO_4$ dissolved in the glass phase, then the amorphous phase must contain

$$z\text{P} + y\text{P} + y\text{Al} \quad \text{gram ions}$$

and

$${}^{g}R = \frac{{}^{g}\text{M}}{{}^{g}\text{P}} = \frac{3{}^{g}\text{Al}}{{}^{g}\text{P}} = \frac{3y}{z + y}$$

Now

$$y = \frac{{}^{g}R}{3 - {}^{g}R} z$$

is equation (3.14) corrected for Al = 3*M*. If $y = 1$, then all $AlPO_4$ is dissolved and the system is totally amorphous.

$${}^{g}R = \frac{3}{z + 1}$$

or

$$z = \frac{3}{{}^{g}R} - 1$$

To calculate the quantity of amorphous material in the aluminum phosphate enamel from equation (3.18),

$$z = \frac{3 - {}^{t}R}{{}^{t}R} \tag{3.27}$$

$$y = \frac{{}^{g}R}{3 - {}^{g}R} z \tag{3.28}$$

Substituting for *z*,

$$y = \frac{{}^{g}R}{3 - {}^{g}R} \times \frac{3 - {}^{t}R}{{}^{t}R} \tag{3.29}$$

Here *y* may be calculated by analyzing the amorphous phase for ${}^{g}R$ while *z* is a known in this case, or it may be obtained from equation (3.27) and ${}^{t}R$. By substituting for *y* and *z* in the following equation the percent crystalline $AlPO_4$ may be easily calculated:

$$\text{wt\% Amorphous} = \frac{121.95\,y + 141.94\,z}{121.95 + 141.94\,z} \times 100 \tag{3.30}$$

where 121.95 is the formula weight of $AlPO_4$.

As with the potassium Kurrol's salt, one must have an analysis of the amorphous phase to obtain ${}^{g}R$. This is more difficult, since the enamel is likely to be very sparingly soluble and it may be difficult to dissolve the glassy phase without dissolving at least a part of the crystalline $AlPO_4$. Perhaps the best estimate can be gained from a measurement of the refractive index of the amorphous system. A Becke line approach would probably work well if a calibration curve is prepared or refractive indices are calculated from molar refractions of Al_2O_3 and P_2O_5. Both approaches have been used satisfactorily in the past. The dispersed phase does not usually influence the refractive index of the continuous phase if it is present in relatively small amounts.

It is now possible to determine if the assumptions at the start of this section are reasonable. They were purely guesses as to what the change would be *if* a certain quantity of $AlPO_4$ crystallized, but in no way considered how much $AlPO_4$ *would* actually crystallize in a real sample. It should be recognized that this is nothing more than a solubility of $AlPO_4$ in a glass. Since solubility cannot be predicted in even an aqueous system, it should not be surprising that it cannot be calculated in a melt system. The refractive index measurement, mentioned above, is simply a measure of the solubility of $AlPO_4$ in the amorphous matrix at room temperature.

From Example 1 above ${}^{t}R = 18/16 = 1.125$, and when the crystallization was completed ${}^{g}R = 16/16 = 1.0$. From equations (3.27), (3.28), and (3.29), $y = 0.833$, $z = 1.667$, and the percent amorphous phase is 94.3% of the weight of the sample; or 5.7% of an enamel of this composition should be crystalline $AlPO_4$. The glassy phase should be very long chain polyphosphates, with n approaching as high as 300 perhaps. This assumes that the system will behave similarly to amorphous alkali metal polyphosphates in this respect.

A similar calculation can be made for Example 2 above. In this case it is anticipated that there should be more crystalline phase in Example 1. The conditions were specified that less crystalline phase should be formed in Example 2 and calculations with equation (3.30) confirm that these were the conditions specified.

3.10. Solubility of Condensed Phosphates

In a strict sense, condensed phosphates have no solubility in *water*. Solubility is an equilibrium quantity of a substance that will dissolve in a specified quantity of liquid at specified pressure and temperature. Most solubility measurements are made at atmospheric pressure if a substance being dissolved is a solid or a liquid. Gases are a special case.

Since all condensed phosphates are unstable with respect to hydrolytic degradation to orthophosphates when exposed to water, a solid or liquid phase is never in equilibrium with a solution phase. Several aqueous phase diagrams of condensed phosphates have been published, including some by the author.[16,17] It was recognized that these studies yielded results that were useful, but not scientifically rigidly correct.

Rates of solution were measured for calcium sodium polyphosphates and calcium polyphosphate fibers. Figure 3.6 is a typical example. Again these results are not strictly reliable because no two systems will yield results that are precisely the same. Variables range from temperature, particle size, rate of stirring, enzymes, impurities, amorous content, influence of grinding or milling, and age of samples studied, to mention but a few. It was not as important to know an exact rate of dissolution as it was to know that samples would indeed dissolve in a matter of

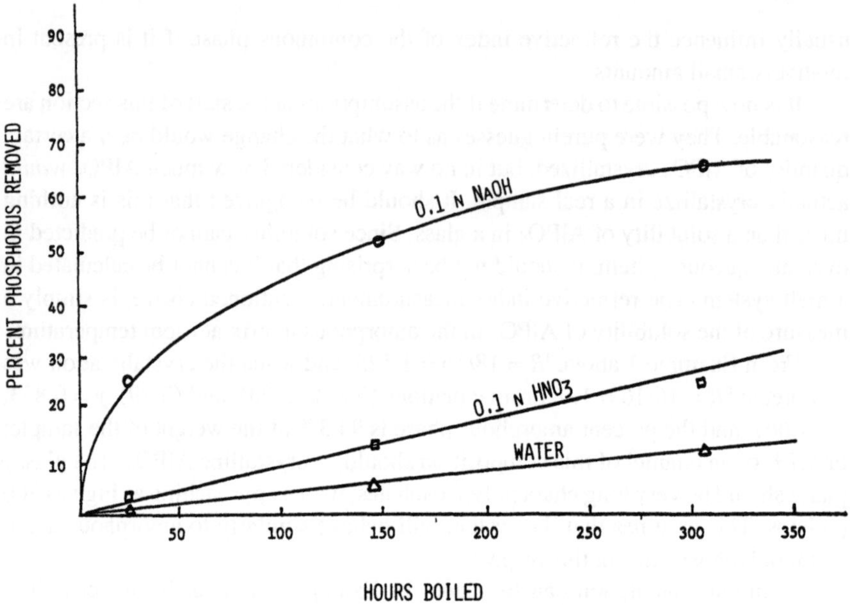

Figure 3.6. Rates of solution of $[Ca(PO_3)_2]_n$ fibers.

weeks rather than decades. This point will be revisited when discussing toxicology in Chapter 9.

3.11. Complexing with Metal Ions

Complex ions and complexes are known by several names. In scientific literature complexing of a cation by a ligan is referred to as complexing. In industrial circles, and particularly in detergent industries, they are called sequestrants, probably derived from a legal term meaning "to seize." When considering polyphosphates, a part of sequestration is probably ion pair formation as well as site binding of cations to a phosphate anion.[18]

Sequestration is an important property of condensed phosphates and their ability to form *soluble* complexes with multivalent cations.[19–21] Many studies have been made of this property. Condensed phosphates, primarily polyphosphates, differ from orthophosphates in the ways they interact with metal ions in aqueous solutions. Orthophosphates remove metal ions from solution by forming insoluble precipitates. They have been used for many years to soften water by precipitating calcium, magnesium, iron, etc. from solutions, but leaving a powdered precipitate that is usually undesirable.

Polyphosphates do not precipitate metal ions from solution, but hold them in solution as soluble complexes, provided the ratio of cations to phosphates does not exceed some value related to a metastable solubility product. It should be obvious that a true solubility product does not exist for the same reasons a true solubility does not exist. If polyphosphates become overwhelmed, a precipitate will form and condensed phosphates function as orthophosphates in removing metal ions from solution.[22]

Rather than attempting to include an in-depth study of complexing agents and stability constants, a general statement will be made and readers are referred to published literature for a more comprehensive view. Perhaps the most perplexing issue about complexing, and equilibrium constants used to study and represent it, is the number of different constants used to represent a system.[23]

Stability constants can be divided into consecutive or stepwise constants, and cumulative or gross constants. Either constant depends upon how the equilibrium equations are written. Consider a stepwise constant first:

(3.31) $$ML_{n-1} + L = ML_n$$

(3.32) $$K_n = \frac{[ML_n]}{[ML_{n-1}][L]}$$

where M is a metal ion, L is a ligand, and K_n is a stability constant.

For no particularly good reason cumulative or gross constants were assigned the symbol β_n:

(3.33) $$M + nL = ML_n$$

(3.34) $$\beta_n = \frac{[ML_n]}{[M]\,[L]^n}$$

Many very clever methods have been developed to measure stability constants. They range from specific ion electrodes to titrations or spectrophotometric measurements that are particularly helpful with colored ions which change color when complexed.

Two of the more important factors in judging effectiveness of a chelant are: How tightly are cations held? What is the capacity of a chelant to hold cations? Both factors are easily observed while titrating a calcium ion solution with a chelant while observing calcium ion activity with a specific ion electrode. The number of milliliters necessary to reach an endpoint inflection is related to the capacity of the chelant, while the difference in the concentration of free calcium ions is a measure of the tightness of the complex. In practice, the molecular weight of a chelant is also important.

In applied chemistry the smaller molecular weights of chelants are usually preferred to larger molecular weights, because the effectiveness per pound is usually better. Since chemicals are usually sold by the pound, it is likely that a low molecular weight chelant will be more cost effective. There are many other considerations when choosing a chelant, but these are important if large volumes are to be purchased or sold.

3.12. Threshold Agents

A lesser known property of polyphosphates is their ability to behave as threshold agents to prevent precipitates from forming. Precipitation is prevented even when concentrations of polyphosphates are too low for them to function effectively as a complexing agent. This property has been used for many years to prevent scales from forming in boilers, hot water heaters, cooling towers, and similar places where scales of calcium carbonate lime, calcium sulfate, or even calcium phosphates could be damaging to equipment. This remarkable property of polyphosphates is very effective. Only a few parts per million concentration are required to keep scales from forming.

Sand filters are used in most municipal water companies to remove parasites (such as liver flukes) from drinking water. Many Asians are infected with liver flukes that attack their livers and are very difficult to eliminate from a patient once they are infected. Untreated, a sand filter will cement grains of sand together in a short time as a result of calcium carbonate and other insolubles "cementing" grains of sand. Once a filter is cemented it is practically useless. But if a water company includes about 100 to 200 parts per million polyphosphate in its potable water, sand filters are effective for several months before their sand must be replaced. It is an area where eutrophication of our water resources has never been an issue.

Phosphate detergents have been removed from detergent markets in the United States since 1992. With the demise of phosphate detergents, condensed phosphates in sewage pipes also will disappear. There was no other source of condensed phosphates sufficient to act as threshold agents. For forty years sewers of the world have been threshold treated with polyphosphates derived as a byproduct of phosphate detergents, for free. This is the same treatment steam boilers and water cooling towers have received for many years to prevent them from scaling and crusting from water hardness.

It will be interesting to learn how much benefit we derived by a constant threshold treatment of our sewers and if it is possible that the plague of zebra mussels could have been prevented if phosphate detergents had not been eliminated. "Green detergents" rely strongly on sodium carbonate, zeolytes, very high dosages of organic surfactants, and sodium silicates for their activity. Not only are these precipitating agents, with orthophosphates derived from urine, but there is no longer a threshold agent to prevent them from cementing sewer lines in much the same way sand filters are cemented if water is not treated with polyphosphates. As precipitation and scaling commences, smaller lines near homes should be the first to notice a difference.

Fifty years ago most homes were equipped with grease traps. These traps were required to collect scaling and lime soap scum preventing sewer lines from clogging. Visits from plumbers to clean grease traps were a regular and expected expense before phosphate detergents were introduced in 1946. Hopefully we shall not be forced to return to grease traps and plumber visits to clean them.

3.13. Hydrolytic Degradation (Hydrolysis)

Of all properties of condensed phosphates, hydrolytic degradation (hydrolysis) has received most attention in the scientific literature. It is an important property of phosphates, but it is also very easily studied. Although the word "hydrolysis" is often loosely applied to degradation of phosphates in much the way it is applied to a reaction of water with an organic ester, this usage is poor. Since phosphates are excellent buffering agents, they react with water as a reversal of neutralization resulting from the slight ionization of water. This reaction is the usual meaning of hydrolysis. A salt of a strong base and a weak acid may react with water to produce a solution with a basic pH. Likewise, a salt of a weak base and a strong acid may react with water to produce a solution that is acidic.

Hydrolytic degradation of polyphosphates is the main theme of this book. Not only is it of ultimate importance (Focal Points 8, 9, 12, and 13) in phosphate chemistry, but also in biochemistry and most other biological sciences. It is the hydrolytic degradation of ATP that has been discussed over and over as an example of the types of reactions expected from phosphate fibers.

It was noted in Chapter 2 that hydrolytic degradation may be represented as:

Hydrolytic Degradation

```
  O O O O                            O O O       O
XOPOPOPOPOX + HOH   ====>   XOPOPOPOH + HOPOX
  O O O O                            O O O       O
  X X X X                            X X X       X
```

while classical hydrolysis may be represented as:

Hydrolysis

```
  O O O O                            O O O
XOPOPOPOPOX + HOH   ====>   XOPOPOPOH + X+ + OH-
  O O O O                            O O O
  X X X X                            X X X
```

Again X may represent one equivalent of almost anything cationic.

Two classes of studies have been published. Most have dealt with the kinetics of degradation, while a few have attempted to define the products of these reactions.

More than 100 articles have been published dealing with hydrolytic degradation. Several quality reviews will be cited rather than repeating the entire list.[24–26]

A minimal understanding of the degradation of long-chain *polyphosphates* is needed in this work. A minimal understanding will not suffice in phosphate areas when all condensed phosphates, including ultra-, meta-, and polyphosphates, are considered. Even the products of degradation of some more recently prepared crystalline ultraphosphates have yet to be defined, not to mention amorphous systems containing a variety of products. Moreover, retardation of hydrolytic degradation is an area that has hardly been addressed. This should be very rewarding both for biological applications and also far more provincial uses of condensed phosphates.

Again, rather than attempting to present a comprehensive coverage of kinetics of hydrolytic degradation of phosphates, I shall present what I believe are some reliable generalizations that can be gleaned from literature data. For those who are interested in more complete data, some very good reviews will be cited. The subject will be addressed for all condensed phosphates. A condensed phosphate is any phosphate containing a POP linkage and therefore is subject to hydrolytic degradation, since all POP linkages are subject to hydrolytic degradation, irrespective of the molecule in which they may reside.

1. All POP linkages undergo hydrolytic degradation.

2. In pure aqueous media the rates of degradation will depend upon the specific phosphate being considered, pH, temperature, and concentration. Hydrogen ions catalyze the rates of conversion of polyphosphates to orthophosphates and shorter phosphates by removing an end group. The rates of degradation of metaphosphates (rings, middle groups) are catalyzed by both hydrogen and hydroxyl ions with a minimum of rate near neutral pH.

3. Most hydrolytic degradation obeys first-order kinetics with respect to the phosphate.

4. Phosphates degrade by several different paths:

(a) Rings of metaphosphates may be opened first to polyphosphate chains and these may degrade as polyphosphates.

(b) Polyphosphates may degrade by at lease two routes. All polyphosphate end groups degrade by the loss of an orthophosphate from the ends of chains, Ge, end groups, leaving a chain one PO_3 group shorter. Polyphosphates with four or more phosphorus atoms per chain may degrade by a reversion to a ring metaphosphate. This is not strictly hydrolytic degradation, because water is not involved in the reaction. A trimetaphosphate is usually formed leaving a polyphosphate that is three, Gm, middle groups, PO_3 groups shorter. It has also been reported that a polyphosphate may split some place in the center portion of a chain leaving two fragments; the sum of the lengths will equal the original molecule. Longer-chain molecules may be subject to this form of scission.

(c) Triply linked groups, Gt, in POP linkages are very susceptible to rapid hydrolytic degradation almost anywhere in a phosphate molecule. Too little is known of this type of degradation to state whether or not stearic hinderance may make a difference in rates. More will be presented on this subject in the discussion of electron density of POP linkages and the rates of degradation.

3.14. Electron Density of a POP Linkage

In this section rapid degradation will be considered unstable in water. This does not necessarily correlate to thermodynamic stability directly in the sense of a free energy change, but hopefully it will be easily read and understood.

The average number of valence shell electrons contributed from both oxygen and phosphorus per POP linkage is a maximum for the pyrophosphate ion, the most hydrolytically stable polyphosphate in basic media. The hydrolytic half-life of pyrophosphate in basic media may be more than two thousand years. The maximum number of electrons that can contribute to a POP linkage stability is found in the pyrophosphate $P_2O_7^{-4}$ ion. Each uncharged oxygen contributes six electrons while the charged oxygens contribute seven electrons. Each phosphorus contributes five electrons for a total of 56 electrons surrounding the POP linkage in the $P_2O_7^{-4}$ ion. When the POP linkage suffers attack by a water molecule, the ion obtains eight more electrons divided between the two resulting orthophosphate groups.

Each middle group, Gm, PO_3^{-1}, contributes 24 electrons to a POP linkage in all polyphosphate ions larger than pyrophosphate. Since all POP linkages in ring metaphosphates are equivalent, all metaphosphates contribute 24 electrons per middle group in either chains or rings. As chains get longer and longer, the electron density per POP linkage approaches 24 (Table 3.2).

Triply linked PO_3 groups are the most electron deficient. In P_4O_{10} there are no charged O atoms and there are six POP linkages all of which are composed of triply

Table 3.2. Electron Density of POP Linkages in Condensed Phosphates

Compound	Average number of valence electrons per POP linkage
Phosphorus pentoxide	13.3 (most reactive)
All ring metaphosphates	24.0
Octaphosphate	28.6
Heptaphosphate	29.3
Hexaphosphate	30.4
Pentaphosphate	32.0
Tetraphosphate	34.7
Triphosphate	40.0
Pyrophosphate	56.0 (least reactive)

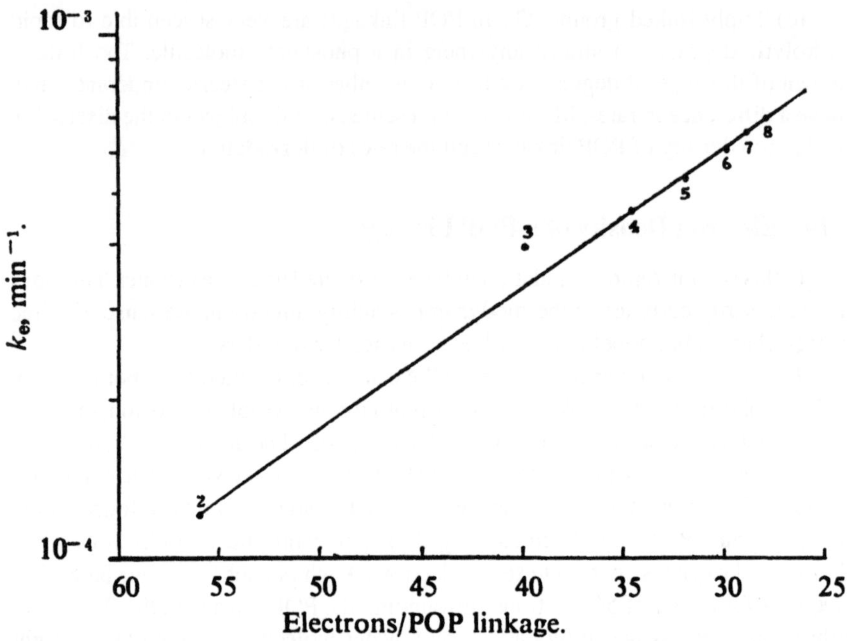

Figure 3.7. Rate constants for end group degradation vs. the average number of electrons in POP linkages.

linked PO_3 groups. There are 13 electrons per POP linkage in P_4O_{10}, the minimum number of electrons of any phosphate and the most hydrolytically unstable. The reaction between P_2O_5 and water is too rapid to obtain reliable rate constants for the common hexagonal crystalline form. Some polymeric forms of P_2O_5 degrade much less rapidly.

Figure 3.7 illustrates a direct relationship between the specific rate constants for the degradation of a polyphosphate at pH 4 and the number of electrons per average POP linkage in various polyphosphates. It is easily seen why the more active bond of adenosine triphosphate is the predominant point of hydrolytic attack of this molecule, since either a hydronium ion or a hydrated cation will be the vehicle of attack.[27]

It is highly significant in phosphate chemistry that the P_4O_{10} molecule is uncharged. This means that the entire charge of any phosphate is contributed by the oxygen in the metal oxide, M_2O, contained in the phosphate molecule. In theory there are no end groups in any phosphate in which the M_2O–P_2O_5 ratio is less than or equal to unity. Because very long chain length polyphosphates can grow in ultraphosphate melts, this is not strictly true. Each chain must terminate in two or more end groups. An isopolyphosphate can have several end groups attached to a single molecule. To the best of my knowledge, these molecules probably exist only by random chance in transient systems and they have never been isolated as crystalline compounds.

The charge associated with any phosphate must lie between zero for P_2O_5 to –3 for orthophosphates. As noted, the charge of a phosphate is contributed exclusively by M_2O or, in the case of acids, by H_2O. Since the thermodynamic stable state is an orthophosphate with charge –3, it is not surprising that the rate of hydrolytic degradation of a condensed phosphate is a function of charge of the molecule. Pyrophosphates are the most stable because they are the most electron saturated of any condensed phosphate.

In any metaphosphate phosphate molecule

$$Gm = P = M$$

In any polyphosphate molecule

$$Ge = 2 \qquad \text{and} \qquad Gm = P\text{-}2$$

It was shown by Griffith and Buxton that the rate of hydrolytic degradation of an end group phosphate is a function of the number of electrons that can influence the charge of a POP linkage in mildly acidic solutions (pH = 4).[28] This is significant in the hydrolytic degradation of condensed phosphates in living systems. The greater the electron density of a POP linkage, the more resistant the POP linkage toward hydrolytic degradation in neutral and basic media, degrading at a slower rate.

O O O

A OPOPOPO_

O_O_O_

H^+

HOH

Let **A** represent the adenosine group. It will be noted that when the hydronium ion, H_3O^{+1}, attacks a POP linkage in adenosine triphosphate, the terminal POP linkage is more negatively charged and the attack usually occurs upon this linkage. Almost any soluble cation that can be hydrated can increase the rate of degradation of a POP linkage, everything else being constant. It is believed that the cations function as a positive carrier of water to direct it into the reaction zone. It was demonstrated many years ago that inorganic polyphosphate degradation obeys the Bronsted–Bjerrum rules with respect to ionic atmospheres.[29,30]

It is recognized that we are discussing two electrical behaviors. In the case of the formal charges, as at the end of a polyphosphate molecule, there is the attraction of a positively charged hydronium ion toward this charged site. The Bronsted–Bjerrum rule is effective under this condition. It is also known that end group degradation is not catalyzed by hydroxyl ions and, if tetramethyl ammonium phosphates are degraded, the rates continue to decrease as the pH is raised toward twelve. The apparent catalysis by strong solutions of NaOH is a result of the hydrated Na^{+1} ions and not the OH^{-1} ions.

The second type of electrical behavior is that related to the average number of electrons associated with the covalent POP linkages. The comparisons of rate in this case were in mildly acidic solutions, where the pH was equal for each polyphosphate studied.

Since phosphatase enzymes are polypeptides or proteins capable of zwitterion behavior, the activity of these enzymes is probably nothing more than their ability to behave as sterically acceptable water carriers that may function as either positively or negatively charged carriers. The reactions between proteins and polyphosphates will be discussed below. The protein's ability to form coascervates with polyphosphates, but not with orthophosphates, is surely significant in the role of enzymes to influence the rates of degradation of condensed phosphates in both biochemical and inorganic polyphosphates (Focal Points 7 and 8). The hydrolytic degradation of polyphosphates has been shown, mostly by some outstanding work of C. Y. Shen, to follow first-order kinetics to very high concentrations, and nonenzymatic reaction rates can become extremely rapid as counter positive ion concentrations are increased. Osterheld reviewed the nonenzymatic degradation in vivid detail and remains a classic article dealing with this subject.[6]

Controlled retardation could be of special benefit to the application of phosphate fibers. In this application of phosphates there is always a compromise between a fiber that is functionally strong and refractory, and one that is harmless because it degrades and is absorbed under conditions in which it serves no useful function. Perhaps a fiber of chrysotile "asbestos" containing weak links, subject to degradation, under conditions where safety was of paramount interest, could be ideal. See Chapter 5 For details.

A chrysotile silicate containing a few hydrolytic degradable groups in each chain would probably be a very desirable and affordable product. Abe's work, revealing a toxic influence when phosphates and silicates are combined in a single molecule, cannot be ignored in this application.[31] If the influence is general, a modified "asbestos" might be more suspect than native "asbestos." As stated earlier, the author is not yet convinced that chrysotile "asbestos" presents a significant hazard to anyone, although it would have been to his advantage to know that chrysotile is extremely hazardous. After extensive research biased against "asbestos," it was not possible to build a case against chrysotile that suggested casual contact with this form of "asbestos" was in any way hazardous.

3.15. Reactions of Proteins and Polyphosphates

One of the early qualitative tests to determine whether or not a substance was an orthophosphate or a condensed phosphate was to mix the phosphate in question with a protein such as egg albumin. If the phosphate reacted with the protein and precipitated, it was likely the phosphate was a condensed phosphate. The text was not foolproof, but was a good indicator. Many polyelectrolytes precipitate proteins. If one knows that the substance under test is a phosphate of some description, then this narrows the room for error.

The gelatin–polyphosphate test is very old and the chemistry is superficially straightforward. Polyphosphate anions react with gelatin cations under acidic conditions. Gelatin is a protein composed of at least eighteen amino acids.[32] Under the conditions of the precipitation test the reaction is probably highly disorganized, but this is not a requirement. It is well recognized that the higher the molecular weights of the phosphates and the proteins, the more efficient the precipitation.[33] Under the proper conditions it is likely that specificity can be brought to the reactions.

In years passed the author engaged in the precipitation of human blood proteins with the sodium tetrametaphosphate ring. With proper care proteins of a highly specific nature can be precipitated. The proteins can be recovered without being denatured when they precipitate with metaphosphates (rings), while the interaction with polyphosphates is very strong and the integrity of the protein is usually violated when removing the phosphate from the proteins.

The interaction of gelatin and polyphosphates is tenacious and the coascervates are refractory enough to form solids that can remain in moist conditions for extended periods. This action is readily observed in the precipitation of gelatin in leather with short-chain polyphosphates.

Dr. Howard J. Buttery of the Encapsulated Products Division of the 3M Company* is a leading scientist in the chemistry of encapsulation. I am indebted to him for the following account of the reactions of gelatin and polyphosphates.

"The term coacervation was first used by Bungenberg de Jong and Kruyt to describe the separation of an aqueous solution of one or more macromolecules into a colloid-rich phase (the coacervate) and a phase of low colloid concentration (the equilibrium liquid).[34] Coacervation and flocculation are closely related phenomena, differing in the higher dispersed state of the latter.

Phase separation can be induced by several processes. Simple coacervation results from introducing an overall water deficiency by the addition of a more hydrophilic agent to a concentrated solution of the macromolecule. Complex

*Dr. Howard J. Buttery, 3M Encapsulated Products, 3M Center Building 235-3E-06, St. Paul, MN 55144, USA.

coacervation is pH dependent and is achieved by charge neutralization of a dilute solution of two or more macromolecules. In certain circumstances both processes can occur simultaneously.

Gelatin coacervates have received the most attention both academically and commercially. However, many other aqueous and nonaqueous systems have been studied for various applications, such as encapsulation.[35,36] Coacervation equilibrium has been applied to fractionation of polymers.[37]

Many biological macromolecules form coacervates, and complex coacervation has been suggested as a model for prebiological organization of primordal organic polymeric polyions.[38]

Conditions necessary for gelatin coacervation can be established by preparing ternary phase diagrams involving gelatin, water, and either desolving agent (simple coacervation) or second macromolecule (complex coacervation). Refractive index and specific gravity measurements have been used to develop such phase diagrams.

Much has been published on ethanol and sodium sulfate for inducing simple coacervation in gelatin solutions, and on the gelatin–gum arabic complex coacervate system. Polyphosphates have been relatively little studied in the literature, but are examples of polyelectrolytes which can participate in both simple and complex coacervate formation.

Encapsulation of hydrophobic liquids and solids is a major commercial application which illustrates coacervation with polyphosphates. First described by Green and Schleicher, simple[39] and complex[40] coacervation can yield capsules for use in controlled drug delivery,[35] carbonless paper,[36] fragrance samplers and cosmetic ingredients.[41] The choice of coacervation system is important, as this can affect the composition of the shell wall and hence capsule properties.

The gelatin–gum arabic complex coacervation process is most favored for encapsulation of fragrance oils. The oil is dispersed in a dilute solution of about equal parts of gelatin and gum arabic held above the gelation temperature at an agitation rate to give oil drops appropriate to the final desired capsule size. The pH is then adjusted to around 4–5 to cause the gelatin to become positively charged. Charge neutralization by the negatively charged gum arabic initiates the formation of coacervate droplets which, upon slow cooling, coat the oil surface and gel to form the capsule shell. The shell contains both gelatin and gum arabic and can be hardened by the addition of cross-linking agents such as formaldehyde or chrome alum for subsequent isolation. For fragrance samplers and carbonless paper, the reaction slurry can be mixed with an adhesive and coated directly.

Essentially the same procedure is reported for encapsulation by gelatin–polyphosphate complex coacervation at pH 4–5.[42] Compared with gum arabic, however, maximum coacervation is achieved at only 1 part of commercially available sodium hexametaphosphates to 10–20 parts of acid-extracted pigskin gelatin. This is due to the lower apparent (ionic) equivalent weight of polyphosphates.[43] The gelatin–

polyphosphate coacervate is reported to have a higher solids content due to higher coacervation intensity, and is predominantly gelatin.

The same procedure at pH 5.5–7.5 using 1 part gelatin and about 3–6 parts hexametaphosphate gives encapsulation by simple coacervation[44] with some polyphosphate being incorporated in the shell."

3.16. References

1. J. R. Van Wazer and E. J. Griffith, *J. Am. Chem. Soc.* **77**, 6140 (1955).
2. R. Pfansteil and R. K. Iler, *J. Am. Chem. Soc.* **74**, 6059 (1952).
3. E. J. Griffith, *Phosphorus Chemistry*, ACS Symposium Series 486, p. 86, ACS, New York (1991).
4. E. J. Griffith, T. M. Ngo, and M. Veiderma, *Proc. Estonian Acad. Sci. Chem.* **42**, 113 (1993).
5. J. R. Van Wazer, *J. Am. Chem. Soc.* **78**, 5709 (1956).
6. K. Osterheld, in *Topics in Phosphorus Chemistry*, Vol. 7 (E. J. Griffith and M. Grayson, eds.), p. 103, Wiley, New York (1972).
7. C. F. Callis, J. R. Van Wazer, and J. S. Metcalf, *J. Am. Chem. Soc.* **77**, 1471 (1955).
8. U. P. Strauss, E. H. Smith, and P. L. Wineman, *J. Am. Chem. Soc.* **75**, 3935 (1953); U. P. Strauss and T. L. Treitler, *J. Am. Chem. Soc.* **7**, 1473 (1955).
9. J. R. Van Wazer, *Phosphorus and its Compounds*, pp. 676, 762, Interscience, New York (1958).
10. N. E. Stahlheber, U.S. Patent 3,723,602 (March 27, 1973).
11. W. A. Filer, Unpublished Monsanto report.
12. T. M. Ngo and F. L. May, Unpublished Monsanto Report prepared for this work.
13. L. Pauling and J. Sherman, *Z. Krist.* **96**, 481 (1937).
14. H. H. Greger, U.S. Patent 2,460,344 (Feb. 1, 1949); U.S. Patent 2,444,347 (June 29, 1948); U.S. Patent 2,405,884 (Aug. 13, 1946).
15. B. Winkhaus, Dissertation, *The Crystalline Similarities Between Aluminum Phosphate and Silicon Dioxide*, University of Bonn (1951).
16. E. J. Griffith and R. L. Buxton, *J. Chem. Eng. Data* **13**, 145 (1968).
17. L. R. Watson and J. S. Metcalf, *J. Chem. Eng. Data* **6**, 331 (1961).
18. M. M. Crutchfield and R. R. Irani, *J. Am. Chem. Soc.* **87**, 2815 (1965).
19. J. R. Van Wazer and C. F. Callis, *Chem. Rev.* **59**, 1011 (1958).
20. A. E. Martell, *Complexing Agents, Encyclopedia of Chemical Technology*, Vol. 6, p. 1, Wiley, New York (1965).
21. M. M. Crutchfield, C. F. Callis, R. R. Irani, and G. C. Roth, *Inorg. Chem.* **1**, 813 (1962).
22. W. E. Brown, *Environmental Phosphorus Handbook* (E. J. Griffith, A. Beeton, J. M. Spencer, and D. T. Mitchell, eds.), p. 203, Wiley, New York (1973).
23. L. G. Sillen and A. E. Martell, *Stability Constants of Metal–Ion Complexes*, The Chemical Society of London, London (1977).
24. C. Y. Shen and F. W. Morgan, in *Environmental Phosphorus Handbook* (E. J. Griffith, A. Beeton, J. M. Spencer, and D. T. Mitchell, eds.), p. 241, Wiley, New York (1973).
25. R. K. Osterheld, in *Topics in Phosphorus Chemistry*, Vol. 7 (E. J. Griffith and M. Grayson, eds.), p. 103, Wiley, New York (1972).
26. J. F. McCullough, J. R. Van Wazer, and E. J. Griffith, *J. Am. Chem. Soc.* **78**, 4528 (1956).
27. J. R. Van Wazer, E. J. Griffith, and J. F. McCullough, *J. Am. Chem. Soc.* **74**, 4977 (1952).
28. E. J. Griffith and R. L. Buxton, *J. Am. Chem. Soc.* **89**, 2884 (1967).
29. J. N. Bronsted, *Z. phys. Chem.* **102**, 169 (1922).
30. N. Bjerrum, *Z. phys. Chem.* **108**, 82 (1924).
31. Y. Abe, H. Hosono, T. Kasuga, and M. Nagase, *Phosphorus Sulfur Silicon* **76**, 243 (1993).

32. J. S. Fruton and S. Simmonds, *General Biochemistry*, p. 125, Wiley, New York (1963).
33. J. R. Van Wazer, *Phosphorus and Its Compounds*, p. 467, Interscience, New York (1958).
34. H. G. Bungenberg de Jong, in *Colloid Science*, Volume 2 (H. R. Kruyt, ed.), Chapters 8 & 10, Elsevier, New York (1949).
35. P. B. Deasy, *Microencapsulation and Related Drug Processes*, Marcel Dekker, New York (1984).
36. A. Kondo, *Microcapsule Processing and Technology*, Marcel Dekker, New York (1979).
37. H. F. Mark, N. C. T. Gaylord, and N. M. Bikales (eds.), *Encyclopedia of Polymer Science and Technology*, Volume 4, *Colloids*, p. 57, Interscience, New York (1966).
38. D. L. Rohlfing and A. I. Oparin (eds.), *Molecular Evolution: Prebiological and Biological*, Plenum Press, New York (1972).
39. B. K. Green and L. Schleicher: U.S. Patent No. 2,800,457 (July 23, 1957).
40. B. K. Green, U.S. Patent No. 2,800,458 (July 23, 1957).
41. H. J. Buttery, in *Cosmetics 1994 Conference Proceedings*, pp. 171–177, Verlag fur chem Industrie, Augsburg (1994).
42. M. V. Fogle and G. Horger, U.S. Patent No. 3,697,437 (October 10, 1972).
43. T. Lenk and C. Thies, in *Coulombic Interactions in Macromolecular Systems* (A. Eisenberg and F. E. Bailey, eds.), ACS Symposium Series 302, pp. 240–250, ACS, Washington DC (1986).
44. G. Horger, U.S. Patent No. 3,872,024 (March 18, 1975).

4

Ropes, Strings, Threads, Fibers, and Molecules

Matter can have many forms and properties. Properties are often exaggerated in systems that are highly asymmetric. Fibers are unique in this respect because they exhibit the maximum asymmetric configuration that solid matter can attain. Despite the gross uniformity of a fiber the average bulk properties must vary across the diameter of the fiber, because the surface and interior must exhibit grossly different structural, chemical, and physical properties. This is true whether the fibers are crystalline or amorphous, natural or synthetic, inorganic or organic, spun or grown from melts or solutions.

Fibers of all types have played an important role in the development of civilization. In this chapter a definition of a fiber will be discussed and why this particular form of matter is unique. Many physical properties of fibers are related to less obvious variables such as asymmetry, surface tension, diameter, length, surface-to-volume ratios, and internal structure. The changes of properties that occur as assemblies of fibers are combined and connected in various configurations as related to phosphate fibers will also be explored.

4.1. Condensed Phosphates

Condensed phosphate molecules come in almost any size and shape, but are always very small compared to most things encountered on Earth. A condensed phosphate may be a polyphosphate (chains), a metaphosphate (rings), or ultraphosphate (branched and cross-linked molecules). A condensed phosphate is a molecule containing at least one POP linkage. Molecules are the smallest chemical compound particles in nature that can have the same properties as pure substances containing them.

Inorganic condensed phosphates may be either acids or salts and are highly ionized. The molecules are sometimes called molecule–ions. The POP linkages are covalent linkages while the cations are held by ionic bonding. Some molecules are tiny spheres, some are hexagons, triangles, or tetrahedra, while other molecules are thread-like or chains. Oils are made of "oily" molecules that slip past each other easily. It is thread-like molecules that will receive most of our attention.

4.1.1. Orthophosphates—M_3PO_4

Orthophosphates are the monomers of the world of phosphate polymers. All condensed phosphates are polymers and orthophosphates are the unique single units. This means that a phosphate is any phosphorus-containing molecule in which the phosphorus atom is contained inside a tetrahedron of oxygen atoms. The oxygen atoms may be replaced by sulfur, selenium, and similar atoms to form thiophosphates, etc. but the phosphorus atom remains caged in the tetrahedron.

In the world of phosphates, orthophosphates and polyphosphates are asymmetric in their properties. Orthophosphates are a single center radical, with sp^3 hybridized bonding, and there are three levels of cationic attraction to the PO_4^{-3} group. This influence is most pronounced with hydrogen ions showing extremely different pKs.

4.1.2. Polyphosphates—$M_{n+2}P_nO_{3n+1}$

Polyphosphate molecules are thread-like and may exhibit extreme asymmetric properties. It is this unbalance of mass and charge that causes them to be the work horses of the phosphate systems. There is little or no practical utility of the other two phosphates systems, meta- and ultraphosphates, except for P_2O_5.

4.1.3. Metaphosphates—$M_nP_nO_{3n}$

The metaphosphates are closed rings and are almost symmetrical in every respect. In solution they behave similarly to any soluble univalent electrolyte rather than a one–three electrolyte and even their relatively high charge does not influence properties that are usually desired in functional molecules. The ring phosphates are hydrolytically degraded slowly in solution and the degradation is catalyzed by both hydrogen and hydroxyl ions. The chain phosphates that result from the degradation behave as any polyphosphate and have many functions.

4.1.4. Ultraphosphates—$M_{nR}P_nO_{n(5+R)/2}$; $R = M_2O/P_2O_5$

Ultraphosphates can be both symmetrical and asymmetrical. There is an all but limitless number of structures that can be prepared, but the properties of most are still unknown. It may be that this group of phosphates will have many new and exciting useful properties if the rapid hydrolytic degradation can be controlled. Phosphorus pentoxide, P_4O_{10}, the end member of the ultraphosphate system, has uses as a dehydrating agent and a very reactive molecule for both organic and inorganic chemistry. It will react with sodium carbonate even in the dry state and can react with lime, CaO, explosively at elevated temperatures. This molecule is highly symmetrical until one triple link is allowed to react.

4.2. Initiating a Project—Strings Attached

When initiating a project to explore possibilities of growing crystals capable of behaving as useful fibers, some simplistic ground rules help in the process of attempting to define first experiments. If a project is truly *new*, often there is little reliable information upon which to base initial trials. New is a relative term, because nothing is completely new. Each experiment is built on the bones of those who laid the foundations for experiments, earlier inventors and scientists. If one had to start each new project with no information other than what he could personally generate, almost nothing "new" would ever be invented.

4.2.1. Inventors

I have worked for, and with, scientists and inventors for more than half a century. It came to my attention long ago that the greatest impediment to most would-be inventors is to overcome an inertia to avoid the first experiment. When this is combined with an innate fear of failure, many wonderful inventions never come to life. Successes one has had in previous endeavors help in attempting bolder experiments while attenuating the fear of failure. The knowledge that even if experiments fail, one can fall back on previous successes, is comforting. I have known this comfort many times in my career.

4.2.2. Approaching a New Project—Untying Knots

After a project is underway and it is known more clearly which natural rules govern the game, more sophisticated questions can be approached. In the early stages a broad brush approach is usually more rewarding. Nature is relatively free to answer "yes" or "no" questions, but is much more demanding and reluctant to answer quantitative questions. However, computers make it abundantly evident that if enough "yes" or "no", "+" or "–", "1" or "0" qualitative questions are asked, answers can become quantitative.[1]

To approach a new project requiring a demonstrable practical end, very general initial approaches will usually help to direct a meticulous painter from painting a Rembrandt on an outhouse. Too often, much wasted effort is spent in detailed studies of a wrong project. In an industrial setting where money dictates project mortality, often only one serious approach is tolerated.

Any number of project probes can usually be launched before a declaration of commitment is required. This is usually called "bootlegging" a project while obtaining some encouraging results before a project is called to the attention of those who control budgets. These persons are seldom swayed by scientific reasons, but can be persuaded by tangible results, particularly if they can gain from a venture.

It is at this stage that the supervisors usually take the project away from the inventors who nursed the project through the "ugly baby stage" that most projects must traverse. If the inventors and technical staff can persuade newly appointed directors (there seems to be no other kind) to ignore the project for a while, the new product may have an even chance of reaching the market.

"Make your boss a hero." This seldom fails and there is usually enough credit to go around. As your boss grows in corporate statue, your boss is unlikely to forget you, if you were helpful in obtaining increased prestige and you provide assurance that you will continue to deliver. It might be that without you, your boss also may fear failure. It has long been recognized that the way to control wild animals is to cause them to become dependent on you. Irrespective of how wild the boss animal may appear to be, if the boss knows that life will become very complicated without you, you are safe. Without you the boss animal might be forced to return to laboratory work rattling beakers.

4.3. Materials and Morphology

If one chooses to make a rope, or a thread, it is unlikely that spheres, cubes, or flat plates should be materials of first choice for this construction. Consider a rope as an initial example. Many ropes are made of cotton fibers. These cotton fibers can be considered tiny ropes or threads themselves. However, these tiny fibers are made of molecules that are also thread-like and can be consider tiny ropes or threads themselves. As an example, cotton molecules are long chains of cellulose. It can be generally stated that fibrous substances are usually, but not always, made of "fiber-like" molecules. A metal wire could be fibrous while its atoms are *not* arranged in a thread-like structure, but most commonly encountered fibers are built on molecular chains.

4.3.1. Fibers Lead to Fibers

If a body is composed of single independent unattached threads, it can be expected that this body will be very flexible, such as a cotton thread or rope. If bridges or cross-links are formed between independent molecular threads, pronounced changes will appear in macrophysical properties of a body composed of these molecules. As an illustration, if two cotton ropes are cross-linked with short strands of rope to make a rope ladder, some degrees of freedom of these ropes are lost.

4.3.2. Bridges between Chains to Form Ladders

A bundle of chain molecules, all of which may slide past each other when stressed, can probably be stretched without breaking, more easily than molecules with bridges between them, as in a rope ladder. This behavior is evident in the stress–strain curves of Young's modulus (see Table 4.1). Everything else being constant, it is be expected that the bridged molecules will produce a fiber that is more difficult to deform than a fiber built of independent chain molecules. An example of this behavior is the change in properties of Nylon filaments when stretched just after being spun, to "crystallize" the fibers. Here, the increase in strength comes from hydrogen bonding rather than formal covalent bonds, but the results are dramatic.

Chrysotile and the amphiboles differ in the manner just mentioned above. The double chains, with formal covalent linkages, not only give the amphiboles more strength than chrysotile but account for the greater chemical resistance and much higher temperatures of decomposition. The serpentine minerals are more inclined to decompose than to melt as such. These minerals can be melted, but some form of phase transition usually occurs before melting.

4.3.3. Bridges between Ladders to Form Nets

If many rope ladders are further cross-linked, with strands of rope between ladders, a net is formed and more degrees of freedom are forfeited. If many nets are cross-linked to other nets to make layers of nets that are connected by short links, a three-dimensional body results from these nets that are nets in three dimensions. This body can become very rigid, particularly if the connecting cross-linkages are short in all directions.

4.3.4. Bridges between Nets to Form Objects

Diamond, the hardest substance known, is an example of a substance where nets are cross-linked to nets on either side with very short cross-linkages. Germanium and silicon are similar substance to diamond, except that cross-linkages are much longer. Silicon and germanium are much softer than diamond. Graphite is diamond where the cross-linkages between nets have been broken and detached nets slide past each other as lubricants. Diamonds are very easily converted to graphite by merely heating them in the absence of air. It is not absolutely necessary to exclude air, but diamond is inclined to burn rather easily.

Each of the above conditions is found in molecules from which fibers may be formed. Polyphosphates and chrysotile "asbestos" are examples of systems that are composed of chain molecules *not* cross-linked to other chain molecules. Likewise, fibers resulting from independent chain molecules are also independent threads. Herein lies one great difference between chains of silicates and chains of phosphates. Cross-linking in condensed phosphates is a triply linked group, highly susceptible to hydrolytic degradation. It is these triply linked groups that make P_2O_5 one of the best desiccants known. These triple links are electron depleted, and are "hungry" to utilize electrons of water molecules and highly energetic.[2]

4.4. Silicates

Silicates can form very stable cross-linked structures that do not suffer hydrolytic degradation except under extreme conditions of pH and\or temperature. Quartz is an example of a tightly cross-linked silicate. Its phosphorus analogue is P_2O_5. Quartz contains a maximum of not only triple linkings, but also quadruple linkings. An SiO_2 group can be attached to four other SiO_2 groups to form a three-dimensional network that extends in all directions similar to diamond, but with longer cross-linkages. Quartz crystals can be stable for centuries while exposed to all manner of environmental abuse. Phosphate analogues would not last one day under similar conditions. Herein lies a significant safety advantage of mineral fibers based upon condensed phosphates. This property can be of value in many types of material science applications.

4.4.1. Amphibole Structures

Amphibole types of "asbestos" are composed of double chains that are linked similarly to the ladder example. These double chains are inclined to produce crystals that may be viewed as ribbons or laves. This is to be expected because ribbon-like molecules can stack most easily to form ribbon-like crystals. Cross-linkages between chains of an amphibole crystal contribute to the stability of amphibole "asbestos" and are responsible for higher temperatures of decomposition and resistance to chemical attack of amphiboles. Chrysotile "asbestos" will quickly degrade in acidic media, while crocidolite, an amphibole which upon casual inspection has an appearance similar to chrysotile, is highly resistant to acid attack.

4.4.2. Crocidolite Filters

Crocidolite "asbestos" was used for many years in analytical chemistry laboratories as a filter medium in Gooch crucibles. Gooch crucibles were small porcelain crucibles with tiny holes in the bottom specifically designed to use crocidolite as a filtering medium that could be heated in a furnace to dry a precipitate before weighing it for analysis. A filter bed was formed by pouring a slurry of crocidolite fibers over the crucible bottom, under vacuum, to form a filter bed. Crucible and filter beds were then heated in a furnace until they attained a constant weight. A precipitate to be analyzed was poured, as a slurry, being retained on the filter bed.

When the contents of these Gooch crucibles were vacuum "dry," the crucibles were again heated to constant weight. An analysis was the difference in weight of a crucible before and after the precipitate was filtered and heated. Chrysotile would never have withstood treatments that crocidolite endured. This is partly because crocidolite contains double chains of an amphibole, while chrysotile does not, a major difference in these "asbestoses," and few other substances other than crocidolite could have been substituted in this application.

Single threads, double threads, or multiple threads of molecules that are attached to each other begin to make some pronounced differences in properties of fibers, as is to be expected. If threads are composed of molecules that are but single strands, not attached to each other by any chemical bonds, it can be expected that these molecules will slip past each other when stressed. In phosphate fibers, when stressed during the measurement of tensile strengths, these molecules flowed past each other when forces of about 300,000 pounds /in^2 were exerted upon them. A crystalline phosphate fiber can elongate and decrease in diameter without breaking.

4.5. Stress–Strain of Fibers

When stressed, ion–molecules in a phosphate fiber resist breaking from two dissimilar bond types. The POP linkages in a molecule are covalent in nature. At the end of molecules reside cationic groups, and these cationic groups hold molecules together through ionic coulombic bonds. The POP bonds are stronger than ionic bonds. There is yet another type of bonding that occurs between the chains of POP linkages. This bonding is also coulombic and cations, usually possessing coordination numbers of six or eight, are attracted to groups in channels between the chains and add strength and rigidity to fibers. Naturally, the charge of these cations makes a significant difference. Small multivalent cations such as aluminum exert a greater influence than alkali metal ions such as sodium or potassium (see Chapter 6).

In addition to an ability of chain molecules to slip past each other in a crystal, the physical morphology of a single molecule must be considered. In some crystals chain molecules are relatively rigid rods. In others they are zigzag rods, some are helixes, while some phosphates are double helixes, much like DNA. As a result of these configurations the molecules have softness and flexibility that is easily seen in their physical properties.

A macrophysical example of cross-linking is easily seen in products made from fiber glass. When fiber glass is made to be used as batting for insulation, it is usually tacked together with a tacking agent. Several organic resins have been used as tacking agents. Phenol-formaldehyde resins have been a favorite. Tacking agents are used to tack fibers together to preserve the integrity of an object being manufactured from fiber glass. Wall and ceiling insulation for homes and buildings is usually tacked, particularly if it is to be applied in mats. Even blown insulation requires tacking to prevent it from compacting on long standing.

4.6. Insulation and Fiber Glass

Insulation, by its very nature, is usually a low-density material and consequently occupies large volumes per pound. In order to transport fiber glass it is economical to compact fiber glass, effectively increasing its bulk density for shipment. When compressed glass reaches its final destination it is desirable for it to spring back to its original volume when compressing forces are removed. Increased volume is required to regain dead air space and thermal properties. This is called an *R* value and is merely the reciprocal of thermal conductivity, in whatever units may be chosen. A larger *R* value of an insulating material denotes less heat to be conducted through the insulator.

When untreated fiber glass is compacted, fibers merely slip past each other when a compacting force is applied to mats of fiber glass. When compressing forces are removed mats will expand slightly, but volumes will not approach original volumes before compaction. However, if these fibers have been tacked together at points where fibers contact each other in an expanded state, each short length of fiber glass becomes a spring when fiber glass mats are compressed. Now when compressing forces are removed, mats of fiber glass regain much of their original volume and insulating qualities. Similar behavior occurs on a molecular scale, causing fibrous molecules to resist sliding past each other.

4.7. Vegetable Fibers

Most vegetation encountered is elongated. Trees, vines, corn stalks and cobs, wheat, grasses, algae, etc. can exhibit aspect ratios (length/diameter) of 100 or more. It is no great surprise that many substances derived from vegetable matter are fibrous. An often encountered example of nature revealing an internal structure of an object by external morphology is the orientation of a seed in a peach. Most people recognize instinctively that an edge of a peach seed is oriented toward the crease in a peach's surface. Yet, more often than not, fruit of vegetation may not be fibrous, while nut shells can be fibrous. A coconut hull is one example.

4.8. Animal Fibers

The animal kingdom is also rich in fibers, ranging from silk to wool to human hair. Bone structure and teeth of most animals and fish are fibrous in structure. Hides, skins, tendons, and other parts are either natural fibers or can be converted into fibers by processing.

Although natural organic fibers are intriguing and interesting, hundreds of books have been written on this subject and too much information is available about these and synthetic organic fibers to rehash these data here. Instead I shall attempt to review data that is not so easily accessible to most casual readers.

4.9. Knots in Fibers and Molecules

The "phosphate fiber," DNA, has rekindled interest in the knotting of strings and ropes. During the time of sailing vessels much time and attention was devoted to tying all manner of complicated knots. Little attention was directed toward a mathematics of knotting. Knotting can be described mathematically through loops and crossings of the "strands," though admittedly these knots can become complex in three dimensions. To overcome this problem knots are analyzed in two dimensions as the shadows of the knots. It has been shown that 12,985 identifiable knots have 13 or fewer crossings. It is reported that only one knot has four crossings and two knots have five crossings.[3]

John W. Alexander was one of the first to develop polynomials capable of describing knots, but these polynomials can become very involved.[4] If two knots have the same polynomial, they are the same even though the knots may at casual glance appear to be very different. More recently Jones developed entirely new polynomials that can be used to describe even more complicated knots based on links and crossings.[5] Whether or not these types of treatment will have any bearing on understanding long-chain polyphosphate behavior is unknown, but there is little doubt that polyphosphate chains can become very twisted and knotted. Perhaps the simpler polynomials will be useful in future research as more is understood about why some of these systems cannot be crystallized.

4.9.1. Crystallization of Phosphates

The problems encountered in attempting to crystallize some polyphosphate melts may well be a knotting of the strands of polyphosphate molecules. It has been adequately demonstrated that the physical behavior of strands of long-chain polyphosphates is highly dependent upon the ionic atmosphere surrounding these molecule–ions.

If long-chain alkali metal polyphosphate molecules are dissolved in dilute solution where ionization of the molecules is more or less complete, negative charges along the backbone of these molecule–ions cause them to stretch out as a rigid-stick model.[6] If, on the other hand, a positive ionic atmosphere surrounds the polyphosphate, as when the phosphates are salted out of solution by the addition of a one–one electrolyte, such as sodium chloride, then polyphosphate anions coil and twist and probably knot. These precipitates are often rubbery, quickly returning to their original shape when deformed by stretching or compressing, provided the force is not excessive or of long endurance.

4.9.2. Crystallization of Polyphosphates from Solution

Long-chain alkali metal polyphosphates seldom, if ever, crystallize when salted out of aqueous solutions. Knotting and coiling is probably a factor in this behavior. The anions are too irregular to fit into a crystal lattice and POP linkages are not being broken and reformed as they are in a melt. As noted earlier, all polyphosphates we tested could be crystallized from aqueous solution as the acridinium salts. Here stearic hindrance, caused by the very large flat cations,

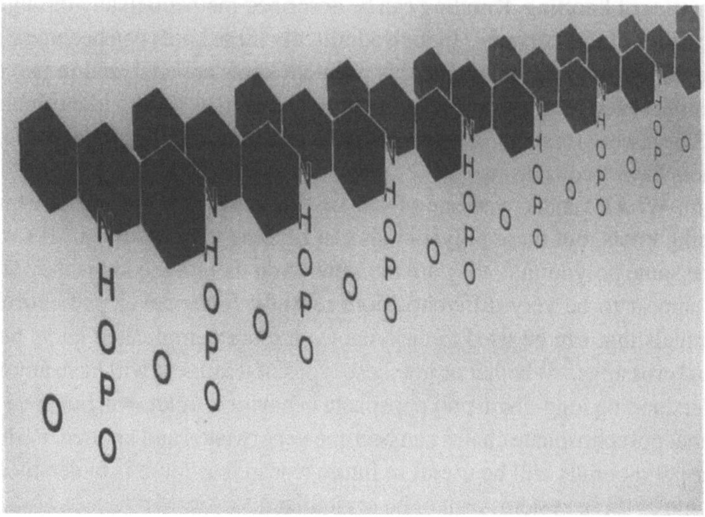

Figure 4.1. Acridenium ions force an alignment of polyphosphate anions to allow them to fit a crystal lattice.

probably prevent these polyphosphate anions from coiling and knotting, thus allowing them to fit into a crystal lattice and crystallize (Figure 4.1).

In discussing hydrolytic degradation, it was noted that in aqueous solutions long-chain polyphosphates could degrade by at least two and perhaps three paths. Two prominent paths are the clipping of end groups to form orthophosphates and a polyphosphate one phosphorus group shorter. The other route, which really has nothing to do with hydrolytic degradation but rather with thermodynamic instability, forms a molecule of trimetaphosphate and a polyphosphate molecule–ion three phosphorus groups shorter. Larger rings have also been observed, but trimetaphosphate is the predominant phosphate formed. This formation of rings could be the result of knotting of the long-chain phosphates, however this is mere speculation and a physical knot may not be required.

4.10. Mineral Fibers

Unlike plant and animal kingdoms, very few mineral substances, such as sand, chert, stones, soil, clays, etc., are naturally fibrous. For this reason it is surprising to learn that properties usually associated with vegetable or animal matter are found in minerals. Moreover, these minerals can have properties so foreign to vegetable matter that they seem almost magical. Most of them will not burn; Most resist high temperatures. Generally, nothing biological or environmental attacks them because they have weathered the ages, exposed to all manner of natural abuse. They are practically insoluble in all common solvents and can have exceedingly small-diameter fibrils with very high strengths. Few manufactured items can match their natural properties.

When discussing these remarkable substances one reaches a juncture of several sciences. Geology, mineralogy, chemistry, physics, and engineering are all prominently involved, but many other disciplines must be, and will be, invoked as needed. What were the conditions that existed to cause a mineral to form very long chain molecules and then to *crystalize* as extremely long slender fibers with aspect ratios of 1000 or more? Perhaps it was the other way around. Maybe two or more molecules joined to start a crystal to grow, creating a template, and long-chain molecules formed as a result of a crystal templet. This question will be explored in detail when considering the formation of synthetic fibrous "minerals."

The attempt to cause a new system to crystallize with a morphology similar to chrysotile is a humbling experience. It is soon recognized that it is beyond all knowledge to cause a second system to crystallize as tiny rolls similar to rolled newspapers, but 200 angstroms in diameter and several centimeters long. All that any experimenter can do is adjust several variables, such as temperature, pressure, electrical or magnetic fields, radiation, composition, seeds, and prayer. If fibers form, it is merely a result of systems that are still too complicated to impose dictated results.

4.11. Risk and Research—More Strings and Knots

Many years ago I was requested to approach a very difficult problem. A few companies had been successfully growing industrial diamonds at high pressures and high temperatures where thermodynamics were known to control these systems. My assignment was to grow diamonds at lower temperatures and pressures by more conventional, organic synthesis methods. I might add that my results were not bad and much successful work today is reasonably close to our approach. However, I soon recognized that it was beyond my power to dictate diamond growth. At best I could adjust conditions and compositions. The ultimate blow to my pride was when I recognized that I was incapable of growing a lowly sodium chloride lattice, if it had not already existed. It is a very complicated structure when considered in this light!

As a footnote, I left this project not because we were failing to make as much progress as I felt we should have been making, but for personal reasons related to potential jeopardy to my family. In the event we were commercially successful. The phosphate fiber project justified my actions.

4.12. Types of Solids

There are only two types of *solid* matter. There are amorphous substances and there are crystalline substances. There are, of course, mixtures of these and there are some substances that have some partial crystalline properties, but should be classed as amorphous, and there are liquid substances that can be classed as crystals.

Amorphous solid substances have no fixed arrangement of their molecular bulks and are not in a definable state, either in energy content or structurally. One view is to consider an amorphous substance as an entanglement of molecules of many shapes and sizes. This is particularly true of phosphate and silicate glasses when considered as supercooled liquids. Their properties depend upon much history. They are not in a thermodynamically defined state. This means that they are not in equilibrium with their environment under existing conditions. Then there must be some other state of matter made of these same molecules (or ions) that has less energy than the amorphous system. These lower-energy systems are usually crystalline.

Because crystals have less energy than corresponding amorphous solids, crystals are stable unless there is another crystalline form of the same substance that has an even lower energy level. Diamond and graphite are both made of carbon atoms, but diamond has a higher energy level than graphite. To convert diamond to graphite is easily done by heating a diamond in the absence of air; the author has done so many times. A trick is to reverse the process, returning lost energy to graphite to form diamond.

An example of phase chemistry is worthy of being reviewed at this time. All transitions from a higher temperature form of a substance to a lower temperature form are exothermic, while the reverse is true when heating a substance that undergoes a transition. It should also be recognized that if a crystalline substance is undergoing a transition from one stable state to a second stable state, this transition occurs at a very precisely defined temperature, a manifestation of stability. For a *pure* substance this temperature can be so well defined that a transition's temperature can serve as a temperature standard. Transitions from ice to water and water to ice are examples in which ice is the substance controlling this system. Without ice, water can be supercooled many degrees before spontaneously solidifying.

Hopefully, it is understood why care must be exercised in choosing a substance from which to begin a project designed to produce fibers that are crystalline and in their lowest energy state under conditions of use. Many times substances can be eliminated without spending valuable time on them. If, from their phase diagrams, it is known that they either have, or are likely to have, phase transitions in the temperature range of application, they can probably be discarded. There are two forms of long-chain calcium polyphosphate crystals. Fortunately, the high-temperature form is but a few degrees below the melting point of this system. A transition under these conditions would not cause trouble because this temperature is already out of the useful range of these fibers. Nonetheless, care must be taken to insure that a high-temperature form has not been prepared by mistake and is awaiting fulfillment, after fibers have been grown.

4.13. Physical Properties of Fibers

When considering physical properties of fibers, it is evident that asymmetric properties must be readdressed. It is unlikely, if not impossible, that a fiber could have properties that were omni-directional. When attracted to an electric field fibers become polarized and stand on end. The molecules within the fibers are usually oriented axially along the fiber. A fiber's tensile strength and impact strengths may differ by orders of magnitude, because cleavage plains within a crystal are oriented at angles to the axis of a crystal. As noted with the $[Ca(PO_3)_2]_n$ crystals, they were inclined to fracture with an angle of 11° leaving their ends very sharp.

The mere fact that a substance may be classed as a fiber does not automatically mean that it is useful. Most known fibers have no known use or very limited uses. Physical properties usually considered in choosing a fiber are strength, durability, and temperature range of usefulness, and also safety if it can be considered a physical property. Under some circumstances hardness will also play a role. In automotive brake systems hardness can become a key factor, because hard fibers can erode metal disks and drums very rapidly. A very large book could be devoted to this subject. Serpentine minerals, "Asbestos," alone had more than 3000 known uses.

Before considering strength, morphology, and heat resistance, durability will be given special consideration, not that it is so important, but it is a vocal focus in modern society. In years before affluent societies were truly affluent, most members did not possess the ability to acquire enough goods and materials to create a trash problem with their discarded "wealth." An idea in manufacturing goods was durability. Today, a product may be judged more on how quickly it degrades, what is not contained in a package or box, or how quickly the product can be recycled. This creates both opportunities and limitations when considering developments of a new fiber, or any other substance for that matter. Phosphate fibers were first envisioned as a new approach to mineral fibers because it was known that they could be rapidly and selectively degraded in biological systems, while maintaining their integrity in most environments protected from water or biological attack.

4.13.1. Chrysotile, $Mg_6[(OH)_8Si_4O_{10}]$, Structure

Chrysotile fibers are composed of threads or tubules that have a structure reminiscent of rolled newspaper or perhaps ancient scrolls of the library at Alexandria. Remarkably, these tubules have diameters of 200 Å. It is difficult to envision how these structures were formed, not to mention what would be required to accomplish this feat with any substance other than magnesium silicate tetrahydrate. These crystals are remarkably resistant to attack by all but the harshest of substances and withstand temperatures as high as 400 °C. This is not a particularly high temperature, but was a safety factor, if one was truly needed, when "asbestos" was used in automobile brakes.

Chrysotile is decomposed at temperatures below those generated on the surface of automobile and truck brakes. Hydrate water is lost at higher temperatures and chrysotile fibers turn to nonfibrous dust, another example where molecular structure dictates crystal morphology. When the shape molecules change, morphology converts from a fiber to a particulate dust with little or no fibrous nature.

Although this author has no reason to admire the asbestos industry, and perhaps has reasons to despise it, I became captivated by the remarkable properties of chrysotile. It is doubtful whether any other substance can begin to fulfill the list of properties exhibited. This is undoubtedly the reason why "asbestos" had so many uses. Not to utilize these properties because chrysotile *may* have some adverse influences would be ludicrous, if this loss were not so serious.

Phosphate fibers were the author's feeble attempt to match chrysotile's properties, in a useable form that must be judged safe by all knowledgeable individuals. However, in the interest of safety there was a sacrifice of properties and there is no hope that phosphate fibers can ever be a one-on-one replacement for "asbestos." Had the author been more knowledgeable of these fantastic properties of "asbestos" when he undertook this project, it is doubtful he should ever have had courage to attempt to challenge chrysotile's dominance, safety or no safety.

4.13.2. Crocidolite, $Na_2Fe_3Fe_2[(OH,F)Si_4O_{11}]_2$, Structure

Whereas fibers of chrysotile resemble rolls of newspapers when viewed as a cross section of fibers, crocidolite is an amphibole and its fibers are laves with a trapezoidal cross section. As mentioned, this behavior is related to the cross-linked double chains of silicate from which these fibers are formed.[7] Like $[NaCa(PO_3)_3]_n$ which is grown as single crystals and must be milled to be converted to fibers, crocidolite is not a natural fiber as it is mined. It must be milled before it is cleaved to fibers. Unmilled crocidolite mineral is a beautiful stone known as either "Cat's eye" or "Tiger's eye." It is also known as Riebeckite Asbestos.[8]

Crocidolite is not unique because it must be milled to become fibrous. Most, if not all, mineral fibers must be either milled, crushed, or stripped to become single fibers. Chrysotile is sometimes found in a condition where the fibers can be easily pulled away from the crystalline spool, in much the same way as $[Ca(PO_3)_2]_n$ fibers may be stripped or pulled from a spool of spherulite that has been leached of its soluble matrix. Figures 4.2 and 4.3 show pictures of chrysotile and our phosphate fiber.

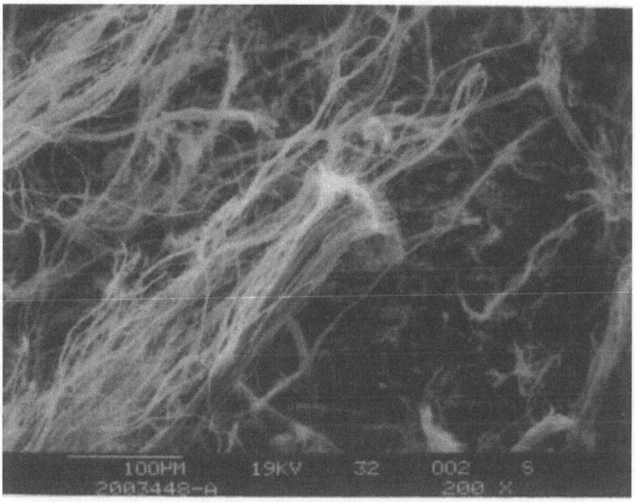

Figure 4.2. Natural form of long fibers of chrysotile "asbestos."

4.13.3. Termolite, $Ca_2Mg_5[(OH,F)Si_4O_{11}]_2$, Structure

Termolite is also an amphibole and its structure is more acicular, tubular, or lamellar. Its crystals are reported to be monoclinic.[8] Termolite is probably the most used fibrous amphibole "asbestos"-type fiber, but crocidolite has been equally popular. Both fibers have been little more than specialty items in the United States with chrysotile fortunately being the primary fiber of choice.

4.13.4. Amosite, $(Mg, Fe)_7[OH\ Si_4O_{11}]_2$, Structure

Amosite is another amphibole that is a very fine needle which can cause copious quantities of dust when handled. It is a mineral that received much attention because of problems it caused for workers in Tyler, Texas and is probably responsible for most of the "asbestos" scare felt around the world. The author readily accepts that this mineral can be very dangerous to work with every day. Again, it is doubtful whether it would cause a problem to anyone on casual contact, although there is some evidence to refute this statement.[7]

4.13.5. Other "Asbestos-Like" Minerals

There are many more minerals that could be discussed ranging from wollastonite, to attapulgite clays, or zeolites. Some mineral zeolites are fibrous "asbestos" types of crystals. Natrolite, $Na_2Al_2Si_3O_{10}{\cdot}2H_2O$, is called Fibrous Zeolite. Zeolites are used extensively in detergents, throughout the world, as a poor substitute for phosphates, but it is unlikely that synthetic zeolites will cause problems as an

Figure 4.3. Form of leached $[Ca(PO_3)_2]_n$ fibers before they are separated.

"asbestos" type of substance. Synthetic zeolites are inside-out fibers with small tubes in crystals that "trap" water hardness cations as calcium and magnesium.

4.13.6. Surface Area of Fibers

By the definition of a fiber in Chapter 1, a maximum fiber diameter (0.5 mm) should yield a surface area of only 80 cm^2 per gram. A fiber of 5 microns diameter yields a surface area of 8000 cm^2 per g, while a fiber of 1μ diameter has a surface area of 40,000 cm^2 per g. Some samples of the serpentine mineral, *chrysotile*, exhibit diameters of fibrils as small as 200 angstroms (500,000 cm^2 surface area per gram). The potential for enormous surfaces from a single 1 g sample of fiber is obvious.

The surface of $[Ca(PO_3)_3]_n$ fibers could be estimated by a treatment of the type evolved above, because they were solid needles with smooth surfaces. $[NaCa(PO_3)_3]_n$ fibers could not be treated in a similar way. They were composed of many tiny fibrils that were inclined to form brushes at the tips. Measured surface areas of $[NaCa(PO_3)_3]_n$ ranged from about 1500 cm^2/g to about 15,000 cm^2/g, depending on how they were milled and the size fraction selected for measurement. The tendency of these fibers to form brushes caused them to have a very high and unexpected oil absorbency.

Dicalcium orthophosphates have been used in mineral oil to form machinery greases for several years. They are superior in this role and most automotive greases are made from the mixture of dicalcium orthophosphate and mineral oils. Although no tests were made of $[NaCa(PO_3)_3]_n$ fibers in greases they should be superior as a cross between dicalcium orthophosphate and "asbestos." "Asbestos" was used as a fiber in lubricating greases for many years before it was found that dicalcium orthophosphate gave better wear protection to moving metal parts in machinery.

Differentiation of equation (1.1)

(1.1) $$S = 2/r$$

yields

(4.1) $$dS/dr = 2/r^2$$

The surface area of 1 cc of continuous fiber increases with the reciprocal of the radius squared. Surface tension is usually reported as γ, in dyne/cm. The circumference, C, of a fiber is

(4.2) $$C = 2\pi r$$

and the cross-sectional area, A, is

(4.3) $$A = \pi r^2$$

If it is assumed that the total strength of a fiber is its surface strength, then

(4.4) $$C/A = 2\pi r/\pi r^2 = 2/r$$

namely

(4.5) $$C = 2A/r$$

Hence

(4.6) $$C\gamma = 2A\gamma/r$$

If F is an equilibrium force that the fiber will support as a result of surface tension, then

(4.7) $$F = \gamma 2\pi r \quad \text{dyne}$$

and

(4.8) $$A = \pi r^2 \qquad \text{cm}^2$$

in which case

(4.9) $$F/A = \gamma 2\pi r/\pi r^2 = 2\gamma/r \quad \text{dyne/cm}^2$$

Tensile strengths of fibers increase with increased surface tensions and decreasing radii. A change of fiber strength as a function of radius can be attributed to a change in fiber surface per unit volume. This has been observed almost universally and highlights the role of surface tension in fiber strengths. This is easily envisioned for amorphous systems, but is more difficult to grasp in crystalline systems. The amorphous systems usually exhibit more symmetrical properties until stretched.

4.13.7. Surface Characteristics of Mineral Fibers

Here, attention will be directed toward crystalline mineral fibers as opposed to amorphous or glassy fibers such as fiber glass or mineral wools. One popular test to which most organic fibers are subjected is heat of wetting. Much can be learned about a fiber's surface when it is learned how much the temperature rises when the dry fiber is contacted with water. Heat is generated because some water molecules contained in liquid water lose their kinetic energy when bound to a fiber. The kinetic energy is converted into heat and, as a consequence, the temperature rises. Much of the difference in the warmth associated with wool as compared to polyester can be attributed to heat of wetting, but this is certainly not the total reason.

As expected, most mineral fibers have a very small heat of wetting when compared to organic fibers, even those that are considered to have low heats of wetting. This tells much about the properties of fibers and what the surface characteristics are on a molecular scale. A large literature has been published dealing with the adsorption isotherms of fibers. These measurements can reveal much about the types of adsorption occurring at the surface of a fiber.

In passing, there are two general types of adsorption on surfaces, physical or van der Waals sorption and activated or chemisorption. In the first case gases may be held to a surface by molecular attraction and there is a relatively low energy of sorption, the order of which is near the energy of vaporization of the adsorbed gas. In the second case the energy of adsorption is relatively large, as though there is a chemical reaction between the gas and the substrate.

Of importance in the case of mineral fibers is the adsorption of substances from a solution in which they are submerged. Willard Gibbs published an isotherm for the adsorption of dissolved substances onto a solid substrate. It was shown that a substance which lowered the surface tension of a liquid at the solid–liquid interface would usually be more concentrated on the surface of a submerged solid than in the body of the solution. These are usually referred to as surface-active agents. On the other hand, substances that raise the surface tension of a liquid in which they are dissolved will be negatively adsorbed at a surface. The concentration of these salts will be greater in the body of the solution than at an interface. In company reports, I have called these salts "body active" salts. Many electrolytes will raise the surface tension of water.

One conceived application of $[NaCa(PO_3)_3]_n$ was as filter media to filter beverages, such as wine or beer. Here, it is imperative that the fiber remove particulates from the medium without removing dissolved ingredients or adding new flavors to the filtrate. New flavors are seldom, if ever, considered an improvement to an established product line. Since the fibers are in every way food grade materials, filter media should have been an ideal application. Spent filter fibers can be utilized as feedstock for animals, with fibers replacing nutrients in much the same way calcium phosphates are used as food supplements.

4.13.8. Useful Temperature Ranges

Useful temperature ranges, like most other topics encountered in applied science, depend upon the needs of a customer and the particular viewpoint of an observer. The immediate reaction to a useful temperature for a mineral fiber is, "How high can the temperature be raised without destroying the fiber's properties?" "How long will the article made from the fiber be serviceable for me if I purchase it?" These may not always be desirable properties, as seen below.

Very few inexpensive substances, particularly fibers, can withstand temperatures that amphibole "asbestos" can endure. However, chrysotile does not withstand particularly high temperatures despite the notion that it does. Since all "asbestos-like" minerals are mined, their properties cannot be precisely specified from one vein of ore to the next. Instead, some general ranges of properties will be specified.

Chrysotile is known to be a tetrahydrate, where water is held as molecular water and not as water of composition. Water of composition, for example, being the type of water formed when two –POH groups combine to form a POP linkage, form one molecule of water that may be vaporized as HOH. There are few crystalline hydrates that will survive temperatures much above 150 °C.

Chrysotile has a useful life to about 450 °C to 500 °C, where the fibers decompose because the crystals undergo a phase transition as a result of "boiling off" water of hydration. Since the surfaces of brake linings run well above this temperature, almost no fibers survive the temperatures generated on a brake drum or disk. On the other hand, the amphiboles are anhydrous crystals and will usually withstand temperatures near 1000 °C, but they are too hard to be of much use in automotive brakes. Chrysotile has a hardness of about 2 on the Mohs hardness scale, meaning that it is a very soft substance, similar to talc (1.5). For comparison tremolite and crocidolite range from about 5.5 to 6.0, making them as hard as the gem stone opal (5.5 to 6.5) and almost as hard as quartz (7.0). It is not difficult to see why they might act much like the semimetallic brake pads that blacken front wheels and wear away brake disks.

4.13.9. Fiber Strength

The first hurdle that phosphate fibers were required to overcome was a tensile strength deficiency, and then a justified harassment by concerned management desiring to know if the fibers had any value for anything, anywhere. The first approach was to calculate the theoretical strength of phosphate fibers while making

Table 4.1. Phosphate Fibers Compared with Other Fibers

	Tensile strength (psi × 10^3)	Young's modulus (psi × 10^3)
Calcium polyphosphate	200	12
Sodium calcium polyphosphate	370	18
Piano wire	350	30
Chrysotile	220–500	25
Fiberglass	200	10
Kevlar	300	12
Nylon	80	0.4
Cotton	60	15
Polyester	35	

a few assumptions. The figure was a respectable 1,000,000 pounds per square inch, but measured values were not this high. The next requirement was to support this value experimentally. Table 4.1 compares tensile strength and Young's modulus of phosphate fibers with several other fiber systems.

The measured values for phosphate fibers compare very favorably with other fiber systems of common use. It is interesting that sodium calcium polyphosphate fibers have a higher tensile strength than calcium polyphosphate fibers. Fiber diameter could account for this difference, or that anions of calcium polyphosphate are more linearly packed in the crystals. Kevlar is a fiber that has been extensively used in "nonasbestos" brakes for automobiles and it performs well.

There were no satisfactory methods to measure tensile strengths of fibers of the type grown in calcium phosphate melts when the Phosphate Fibers Project was initiated. Here again, the genius of Dr. Hinkebein came forward with an ingenious solution to our problem. He built a micro-unit to measure Young's modulus, tensile strength, and modulus of elasticity on single fibers of calcium polyphosphate.

A photograph of Dr. Hinkebein's unit is seen in Figure 4.4. It has a microscope, cemented hooks to attach single fibers with glue, a strain gauge, and a recorder. An analyst can observe the fiber as a stress–strain curve is generated. Dr. Hinkebein confirmed that the calcium polyphosphate fibers were as strong as theoretical calculations had predicted. This strength was also exhibited in work with plastic composites, where these fibers were used for reinforcement.

4.13.10. Chemical Properties of Phosphate Fibers

It is one thing to grow safe fibers, but quite another to know whether or not they are useful for anything other than occupying space safely. The language of materials science is foreign to most chemists of the type found in an industrial chemicals laboratory. Their work is devoted to food products, detergents, fertilizers,

Figure 4.4. Dr. Hinkebein's instrument designed to measure the tensile strength of fibers.

deflocculants, buffering agents, etc. Everyone working with fibers required himself to learn a language seldom used since graduate school. Some few were conversant with terms such as tensile strength, but most still thought in units as pounds per square inch rather than newtons per square meter. Pascals were not to be found in any handbooks usually kept in the laboratory.

On the other side of this coin the group was well versed in concepts such as phase chemistry, solubility, enzyme activity, hydrolytic degradation, and many fields where fibers were utilized to make products. It was the engineering and physics of fibers that were weak, in the initial stages of the work. After a few weeks we were conversing with the more sophisticated scientists and engineers in the areas that needed this cooperation. Before the project was halted some members were contributing new knowledge to areas that had been unknown to them when the project started.

4.13.11. Fibers and Health

In Chapter 9 the toxicology of phosphate fibers and other fibers will be discussed in detail as related to animal testing. Is there some unique property of fibers that causes them to be more of a health hazard than the same chemical, but exhibiting a nonfibrous morphology? There are several physical properties worthy of consideration. The electrical charge distribution of a fiber is dipolar and is concentrated on the ends of the fiber. If the ends of the fiber fray to smaller fibrils, this property can increase the total charge by a capacitor influence as well as increasing the total surface area of the fibers. The charge on a small fiber will attempt to align a fiber with respect to neighboring fibers. This behavior could induce lumping of fibers in lungs, a property that is believed to be harmful when fibers, suspended in liquid media, are injected into the trachea of test animals.

The tendency of small particles to become air-borne to produce dust is always a matter of concern when working with solid powders. The hazards can range from health concerns to dust explosions. Unfortunately, this is another subject that cannot be given a proper treatment in this book. Those with a serious interest should refer to A. D. Zimon's outstanding treatise, *Adhesion of Dusts and Powders*.[9] Some more obvious properties will be noted.

4.13.12. Dying Mineral Fibers

Very little work was done in an attempt to color phosphate fibers because the project did not reach the application stage of making fabrics from either $[Na(PO_3)]_n$ or $[NaCa(PO_3)_3]_n$ yarns. There is small doubt that the crystals could be dyed almost any chosen color. The behavior of organic dyes on crystal surfaces has been highly developed and colloidal pigments of both organic and inorganic substances are well known.

4.13.13. Electrical Properties (Dipoles)

The electrical dipole of the crystalline phosphates caused them to be of interest as potential anticaking and antistatic agents. It was relatively easy to induce a dipole into a crystal with a smaller diameter and a respectable aspect ratio of about 50. This aspect ratio is small when compared to many mineral fibers, but phosphate fibers could be dusted on charged surfaces to kill charge on a surface. Some crude experiments with charged sheets and cloth indicated that the fibers might be useful to counteract static electrical charges.

All crystalline phosphate fibers considered in the Phosphate Fibers Project were probably piezoelectric. Although all were not tested for piezoelectric behavior, all were birefringent when viewed in a polarized field of a petrographic microscope. Few, if any, phosphate crystals that are birefringent are not also piezoelectric.

4.14. References

1. E. J. Griffith, *Caking of Particular Solids*, VCH Publ. Co., New York (1992).
2. E. J. Griffith and R. L. Buxton, *J. Am. Chem. Soc.* **89**, 2884 (1967).
3. I. Peterson, *Science News* **133**, 326 (1988); W. B. R. Lickorish and K. C. Millett, *Math. Magazine* **61**, 3 (1988).
4. J. W. Alexander, *Trans. Am. Math. Soc.* **30**, 275 (1928).
5. V. F. R. Jones, *Bull. Am. Math. Soc.* **12**, 103 (1985).

6. J. R. Van Wazer, M. Goldstein, and E. Faber, *J. Am. Chem. Soc.* **74**, 4977 (1952).
7. W. Simpson, in *Asbestos—Properties, Applications and Hazards*, Vol. II (S. S. Chissick and R. Derricott, eds.), Wiley, New York (1983).
8. J. Bauer, *A Field Guide to Minerals, Rocks, and Precious Stones*, Cathey Books, London (1976).
9. A. D. Zimon, *Adhesion of Dusts and Powders*, Consultants Bureau, New York (1980).

5

Melts and Phases

The first requirement for anyone wishing to make significant contributions to the science of new condensed inorganic phosphates is some understanding of phase chemistry. It is not required that phase chemistry be an end in itself, as with those who are interested in the science of phase thermodynamics as an independent study. Some treatises dealing with phase chemistry are all but incomprehensible to those of us who are interested in using diagrams as tools, but not a life's study.[1] Although I have used and studied phase diagrams for most of a lifetime, I make no pretense of being an accomplished master of all aspects of this science. In phosphate chemistry, as I practice it, phase chemistry is an indispensable tool.

The phase diagrams usually generated in phosphate melt chemistry are simple two- and three-component systems and are relatively easily understood. The aqueous phase diagrams of orthophosphates may contain four or more components and often require special techniques to represent the data. There are no *bona fide* aqueous phase diagrams of condensed phosphates, despite the fact that others and I have published diagrams. Equilibrium is never truly established in these systems because all condensed phosphates are continuously degrading in aqueous solutions. In the Phosphate Fibers Program these systems were hardly ever considered and the aqueous diagrams will be neglected here.

In this chapter melts and phases will be reviewed in theory and in practice. A review of techniques used to study phase science in phosphate chemistry will be developed. A section will be devoted to methods of using published diagrams to predict the behavior of systems that have never been studied. Thermodynamic treatments will be avoided except in areas of solid–solid phase transitions, where they will be briefly discussed.

5.1. Biodegradable "Asbestos"—A Research Project

Phosphate phase chemistry has been a favorite subject of several diverse scientific groups. In silicate glass science, phosphates have been particularly helpful in obtaining a better understanding of silicate chemistry. This is true as phosphates related to geochemistry.[2] In many ways phosphates are more easily studied than silicates and yet they are similar.

It is the similarity between silicates and phosphates that suggested that phosphorus might be substituted for silicon in molecules of serpentine mineral types. If properties of silicates can be retained while adding known safety properties of phosphates, a safe "asbestos"-type product is possible. An early unsuccessful attempt was made by the author to prepare a "safe asbestos" by inserting phosphate groups within silicate chains of chrysotile. With some additional work this should still be possible and no more than two or three percent of the weight of the chrysotile product would be required as inserted phosphates, and probably inexpensively.

It is well known in phosphate and silicate chemistry that heteropoly bonds such as POSi, POS, and POAs are less stable than POP linkages, being highly susceptible to hydrolytic degradation.[3–5] If a PO_3 group were inserted into molecules of the type found in chrysotile, these molecules probably would degrade when subjected to moisture for extended periods. From a safety viewpoint this would be ideal, because fragments of degradation would likely be either soluble or nonfibrous powders.

$$\begin{array}{l} O^-\ O^-\ O^-\ O^-\ O^-\ O^-O^-\ O^-\ O^-\ O^- \\ SiOSiOSiOSiOSiOPOSiOSiOSiOSiO^- \\ O_-\ O_-\ O_-\ O_-\ O_-\quad\ \ O_-\ O_-\ O_-\ O_- \end{array}$$

In this example note that a PO_3 group has been spliced into a silicate molecule. The trick that will need to be perfected is to sneak PO_3 groups into place without destroying integrity or strength of chrysotile crystals. Points of scission, when these fibers are exposed to moisture for extended periods, should be at sites of inserted PO_3 groups and could happen either to the left or right of phosphate groups. Remember that silicates do not undergo degradations of this type and therefore this would give an added new property to chrysotile when accomplished.

In a "head-on" approach to this project, orthophosphoric acid, as dilute aqueous solutions, was added to samples of chrysotile. Next, treated fibers were heated in a furnace to temperatures slightly above temperatures expected to cause formation of heteropoly bridges. A problem encountered in this short-lived project was a detrimental influence of acidity on fiber integrity. Chrysotile is not nearly so resistant to acid as are amphiboles. Other approaches would have probably been more fruitful, but polyphosphates were behaving splendidly as fibers and all attention was directed to them instead of attempting to change existing properties of natural fibers.

This project is worthy of more attention and probably could be accomplished with rather modest resources. Both chrysotile and amphiboles should be considered by anyone choosing to explore this approach. A safe amphibole could be even more desirable than chrysotile. It is possible that a cross-linkage of an amphibole's chains could cause an entire crystal to "unzip" to a harmless powder when subjected to moisture for controlled periods. If –POPOPOP– linkages could be incorporated into a chain, it is even possible they could be made to be "biodegradable," as happens with chain phosphates.

5.2. P_4O_{10} (P_2O_5), The Mother of All Condensed Phosphates

Depolymerization of phosphates is unidirectional in aqueous environments, with the exception of some biological systems that convert ADP and orthophosphates back to ATP, etc. Soluble silicates can be polymerized to silica gels or SiO_2 from aqueous solutions, while sparingly soluble (insoluble) silicates are usually highly refractory. These behaviors of silicates can complicate a study of these systems compared to similar phosphate systems.

If strong phosphate fibers are to be obtained it is necessary to obtain polyphosphate chains of maximum length. We limited our work to two approaches to growing long-chain polyphosphates. In theory, any condensed inorganic phosphate can be prepared directly from P_2O_5 and the corresponding metal oxide. Sufficient heat is liberated from reactions of P_2O_5 and metal oxides to form melts of many systems. A very interesting reaction can be initiated by mixing dry sodium carbonate and dry P_2O_5 and then adding one drop of water to the mixture. Active CaO and P_2O_5 can react energetically enough to explode.

$$P_2O_5 + 3M_2O = \text{Orthophosphates}$$

$$P_2O_5 + 2 \text{ to } 1\ M_2O = \text{Polyphosphates}$$

$$P_2O_5 + 1M_2O = \text{Metaphosphates}$$

$$P_2O_5 + <1\ M_2O = \text{Ultraphosphates}$$

Although a process can be perfected, based upon P_2O_5, this process is very expensive unless P_2O_5 is made directly by burning elemental phosphorus as a step in a process. P_2O_5 is exceedingly hygroscopic, enough to be an excellent desiccant, and particles of P_2O_5 formed by burning elemental phosphorus are so small that they are difficult to capture. These two properties cause P_2O_5 to be an unattractive raw material in most processes where the very fine particles must be captured in a solid state, packaged, and shipped to a location of application. If, however, a process is designed to accept P_2O_5 directly from a small phosphorus burner, the price can become very attractive. The process can be simple. Phosphorus was burned to P_2O_5 by seamen to lay smoke screens for naval vessels before titanium tetrachloride was commercially available. These systems were trouble-free and effective.

5.3. Melt Chemistry

A second approach is to use a more conventional melt chemistry to grow fibers. Conventional high-temperature phosphate melts are not easily contained. Melts of condensed phosphates have been called the Universal Solvent. There are few if any substances they will not dissolve in phosphate melts at higher temperatures.

In laboratory studies platinum, gold, or some forms of ceramics can be used to contain phosphates while they are heated to melts. Platinum is less attacked by basic melts, while gold is superior for melts rich in P_2O_5. In all cases at temperatures above 300 °C crucible corrosion contaminates products to a greater or lesser extent, but contamination can very often be localized to areas of contact between crucibles and melts. Larger melts have smaller surface-to-volume ratios. Purer products are manufactured in these cases. This is one reason why a compound can very often be obtained with higher purity from an industrial process than it can be prepared in a laboratory. This is particularly true with compounds such as $Na_5P_3O_{10}$, which is prepared in large volumes industrially.

Melts Are Difficult To Contain

Phosphate melts are avoided in industrial processes whenever possible, but when very pure products are required from a melt, a frozen lining of the phosphate being prepared substantially reduces contamination derived by corroding and dissolving a crucible into a melt. Melts are contained in a crucible that is made from the frozen melt itself. A crucible is first heated with feed material to be prepared and contained in the crucible. When a melt is formed, most melt is poured out leaving a frozen lining of product as lining in the crucible. This crucible is then recharged with fresh raw materials. Only central portions of this crucible are heated to melt temperatures, while the lining is cooled either by air or by liquid heat exchangers. A flame is usually impinged upon the central feed material. Obviously, this approach cannot be easily applied to small melts of a few grams, but can be easily done with samples of a few pounds.

5.4. Phase Chemistry

Phase chemistry was an established part of the classical physical chemistry curricula in past years. In recent years, less and less attention has been directed to this area because it is not as directly applicable to biological sciences as it is to salt chemistry, geochemistry, metallurgy, ceramics, mineralogy, etc. Although phase chemistry can be very involved if solids, liquids, and gases are considered, in phosphate and silicate chemistry it can be applied rather simply in most instances. We shall consider melting points and solid–solid transitions in this work. Samples of known composition will be compounded, to borrow a phrase from pharmacy, and then melting points of samples will be determined, while noting any other phase changes that may occur along the way.

5.4.1. Phase Diagrams

A phase diagram is a graphical representation of changes that occur in samples of defined composition as a function of temperature and sometimes pressure. Composition is changed from sample to sample. In phase chemistry only three variables are considered: temperature (*T*), pressure (*P*), and composition (*C*). Phase diagrams are defined as equilibrium diagrams, namely, if no outside parameters are changed then the systems studied will remain fixed indefinitely. This is an idealized case and is seldom, if ever, attained in real systems. Nevertheless, an approach to equilibrium is close enough for these diagrams to be extremely useful. In our case they will be little more than temperature–composition diagrams and, with a few minutes reflection on these diagrams, they should be easily understood even by the uninitiated.

Without phase diagrams, condensed phosphate chemistry would be almost impossible to understand. Properly prepared, a phase diagram is an equilibrium diagram, but in industrial processes equilibrium is seldom, if ever, achieved. Often, systems are very slow to reach demonstrable equilibrium. Industrial processes frequently require the production of hundreds of tons of materials in a short period and equilibrium must be compromised. However, if quality crystals are to be grown, even from industrial melts, equilibrium must be approached. This means that melts must be of uniform homogeneity and cooled at a rate that will allow diffusion to crystal surfaces to be orderly rather than forced. For most phosphate systems much reorganization is required for molecules to acquire dimensions and orientations to fit within a crystal lattice.

It is agreed that a phase diagram must be an equilibrium diagram. Without this constraint nothing about these diagrams is reliable. All too often solid products cake badly because a product was packaged and shipped before equilibrium was established. There are times when two states of a system are different, but of very similar energy levels. Under these conditions phase diagrams have been published where one diagram is slightly metastable with respect to the other. These systems are well known for causing problems.[6]

Depending upon the experimenter several techniques have been employed to develop phase diagrams. In phosphate science the late G. W. Morey's technique was probably the most reliable, but very time-consuming. It will be seen that even he could miss a compound in a diagram.[7,8] A compound missed in the original published phase diagram ultimately became the phosphate fiber, $[NaCa(PO_3)_3]_n$. This is no reflection on his work. There is no way to test whether or not all compounds capable of crystallizing in a system have indeed crystallized.

5.4.2. Thermal Studies

Morey prepared his samples for thermal studies by first making certain that they were homogeneous. He did this by precrystallizing samples and then grinding them to a powder. In some cases this action was repeated several times. A carefully weighed sample was then wrapped in a small disk of platinum foil about 1.5 cm in diameter. A pinhole was punched in the foil through which a wire could be threaded. This foil was suspended by a very thin platinum wire contained in a perpendicular electric tube furnace that was itself wired with larger-diameter platinum coils, used as heating elements. Samples were heated to a specified temperature and held for extended times. When sufficient time had passed for samples to reach equilibrium, they were suddenly dropped from the furnace into cold liquid mercury. This was done by passing an instantaneous high current through the fine wire used to suspend the sample of foil. In a flash samples were quenched in cold mercury placed below his furnace tube.

Next, samples were observed with a petrographic microscope. If crystals were observed in the sample at the microscope stage, the furnace temperature was increased and measurements repeated. If no crystals were observed, furnace temperatures were decreased and the measurement repeated. Slowly and methodically, temperatures were adjusted until the last tiny crystal almost disappeared. In this way an equilibrium temperature was established with very high reliability. Although this process was slow and laborious there were probably no more reliable results ever obtained on phase systems. It was, however, difficult at times to detect solid–solid phase transformations solely from these furnace observations. As seen below, differential thermal analyses (DTAs) do not suffer from this limitation.

It should be mentioned that it is possible to determine solid–solid phase transitions by the Morey technique. In this case solids are heated for extended periods above and below a transition temperature and identification of a crystal form may be made either by X-ray analysis or by microscopic examination, as with a Becke line refractive index analysis. Once the refractive indexes of the two crystal forms are established, walking in on a transition can be done in much the same way a melting point can be established.

Yet another way of studying phase transitions and melting points is with a hot stage microscope. A phase transition often can be observed directly with a petrographic microscope. This is particularly true if the crystals are birefringent. When a crystal undergoes a solid–solid phase transition, the angles of extinction of the mother crystal and the daughter crystals are not oriented in the same direction with respect to the polarized field. Hot stages for X-ray stages can be used in much the same way to measure the temperature of a transition.

In other cases when an inert liquid can be found and the temperature of a solid–solid phase transition is not extreme, a dilatometer approach can be used. A change in volume of a solid at a phase transition is observed. Usually, a record of the volume of the liquid system is plotted as a function of temperature. At a phase transition the change of volume is rapid at the transition temperature.

5.4.3. Differential Thermal Analyses

Often, differential thermal analyses are employed to determine phase diagrams. These analyses can be viewed both positively and negatively. A positive view is that thermal behavior, such as phase transitions, can be viewed directly on charts and this method is usually rapid. If samples are heated or cooled too slowly or too rapidly, phase transitions can be passed over without causing changes in a temperature curve. If heated or cooled too slowly, the thermal activity was spread over too large a time span to be recorded. If heated or cooled too rapidly, a transition of the solid itself can be skipped. This is independent of the sensitivity of the instrument.

A negative view of differential thermal analyses is that samples can never be at equilibrium, because a sample is intentionally forced to attain thermal activity great enough to be observed. At very slow heating rates energies are smoothed, and thermal activity is often difficult to detect. At equilibrium, even a very slow approach to equilibrium cannot be detected with most instruments. This causes the uninitiated too often to increase rates of cooling or heating to an extent that supercooling or superheating can cause difficulties. With phosphates and most other glass formers, heating curves are more reliable than cooling curves, because systems are less inclined to superheat than they are to supercool. However, heating curves can be determined at such fast rates that samples cannot keep pace with heating rates.

The author has chosen to analyze larger rather than smaller samples when preparing phase diagrams from differential thermal data. With very small samples it is difficult to choose a sample that is truly representative of a substance being analyzed. Again, surface-to-volume ratios are much larger when using small samples, offering more opportunity for a sample to become contaminated. This is particularly true if a sample is inclined to attack a container. It is sometimes difficult to determine whether or not an observed inflection is a reaction between containers and samples, or is a part of a system under study. In addition, larger samples determined at slower heating rates generate sufficient energy changes to be easily detected and correlated with the other samples used to construct a diagram. Nonetheless, even under the best conditions, differential thermal analysis is not an *equilibrium* method.

Yet another problem likely to be encountered in utilizing differential thermal analyses is identifying a spurious exotherm or endotherm that is real, but does not seem to fit with other samples of this system. More often than not, the author was unable to identify a cause for these thermal activities. It was usually assumed that perhaps a metastable phase with a limited range of stability had entered a system. There was a chance that an amorphous phase was present in small variable quantities and was not detected by any tools used to certify samples. It was to be expected that there would be some behavior that was never satisfactorily explained, within constraints of time and money imposed in industrial laboratories.

5.4.4. Gibbs's Phase Rule

Perhaps too much ado has been made over the Phase Rule. Its beauty is its simplicity in defining the relationship between components, variables, and phases. It is almost self-evident in simple systems of one, two, and three components as usually encountered in phosphate chemistry. The variables of state, temperature, pressure, and composition, are controlled when determining phase diagrams. Pressure is usually atmospheric, temperature may be an active or an inactive variable, while composition is usually defined by either adjustment or measurement. These can define the degrees of freedom of a system so often dreaded by students of thermodynamics. Degrees of freedom are merely the number of variables of state that can be changed without destroying or creating a phase.

Rather than pursuing this subject in depth while creating even more confusion, the reader is referred to almost any good classical physical chemistry text of thirty years ago. To solve the mathematics of the Phase Rule no computer is required; they are usually simple and straightforward.

5.4.5. Industrial Projects

In most industrial laboratories it is a requirement that research efforts yield reliable results, but that lace and trim details be forfeited to more precise scientists! In our laboratory we referred to our research as "Fist Order." It seldom caused problems, but occasionally we were guilty of overlooking an important detail. These were always brought to our attention. Sometimes, we were aware that issues had to be ignored, while sometimes, these details had never surfaced in our work. New details generated by others seldom modified conclusions drawn from our research. Improved understanding of systems contributed by others was always appreciated, even when we recognized that we had overlooked their important contributions.

5.4.6. Crystals of Elemental Phosphorus

In solid–melt diagrams, it is necessary that solid phases be crystalline and that these crystals possess a stable phase under the conditions of a diagram. Some substances can exist in several crystalline states, while their properties are hardly recognizable as the same chemical substances. Elemental phosphorus is a very good example. There is white phosphorus, red phosphorus, and black phosphorus.

White phosphorus resembles water when melted and paraffin wax when frozen. There is no substance that can properly be named yellow phosphorus. Yellow phosphorus is impure white phosphorus. White phosphorus is highly crystalline, pyrophoric, spontaneously igniting when exposed to air. A yellow discoloration is usually caused by a variable substance affectionately named LOOP, and is sometimes called "yallar stuff." LOOP stands for lower oxides of phosphorus. When phosphorus is partially oxidized these compounds are formed.

Red phosphorus is either crystalline or amorphous and exhibits no tendency to ignite when exposed to air at temperatures below 50 °C. While white phosphorus melts sharply at 44.1 °C, red phosphorus does not melt cleanly. Its melting temperature under pressure is high, being over 400 °C, but its exact value is difficult to determine.

Black phosphorus is difficult to obtain, but is similar to graphite in many of its properties. Bridgeman, renowned for his high-pressure work, was the first to prepare black phosphorus.[9] The atoms of phosphorus in black phosphorus are arranged in sheets in crystals that are themselves sheets or "flakes." This is another example of molecular structure expressing itself in particle morphology and physical properties of a crystalline substance. These black phosphorus crystals are difficult to ignite. Crystalline black phosphorus (there is also an amorphous variety) is reported to be a semiconductor of electricity, about 500 times more resistive than graphite.[10]

5.5. Phase Transitions

When preparing synthetic fibers it is usually preferable to avoid crystals that can undergo phase transitions in the temperature ranges of their intended application. This is one reason why amorphous inorganic phosphate fibers may not function well for an extended period. Attempts to prepare phosphate glass fibers have always met with failure in past years *when refractory requirements were desired.* Additionally, molecular weights in glasses are relatively small when compared to crystals of a similar composition. Two problems have usually been encountered.

1. Amorphous solids usually react with water vapor, causing destruction of POP linkages.
2. Since glasses are thermodynamically unstable with respect to crystalline phases, given sufficient time they crystallize (devitrify).

In both instances fibers lose their strength and usually disintegrate to powders. Therefore, if a condensed phosphate is to be chosen for its lasting physical properties, the chances of satisfactory results are better if crystalline fibers can be grown, rather than attempting to spin amorphous fibers.

A spun amorphous phosphate fiber could be of use in areas where it is desired that the fiber degrades rapidly, but adds green strength to a reinforced item. In certain bone prostheses this could be highly desirable. In other areas, where a soluble strong fiber that is biocompatable is needed, as in sutures, spun phosphate fibers should be ideal. Another use is in the weaving of fabric where selective portions are to be removed for either texture or beauty; spun soluble phosphate fibers should be ideal.

5.6. Phase Diagram Systems

One-, two-, and three-component systems will be considered. A one-component system is one substance that defines composition uniquely throughout a diagram. Temperature and/or pressure are usually the variables under consideration to define the state of a system. In phase transition work, variables such as density, heat capacity, optical properties, etc. may be measured to determine changes in phase state as the variables of state are changed. If there are no transitions in a one-component system in a region under consideration, there is very limited phase chemistry as such.

The usual two-component diagram is plotted on rectangular coordinate graph paper. Temperature is usually the vertical axis while the relative percents of the components are plotted along the horizontal axis. The components may be molecules, salts, or ions, depending upon the system under study. In inorganic phosphate chemistry, a metal oxide is usually chosen as one component while P_2O_5 is chosen as a second component.

Three-component diagrams can usually be plotted on a triangular graph with temperature plotted on an axis that is perpendicular to the base plane of the composition diagram. Then the diagram is usually represented as a surface. Usually, P_2O_5 is one component while two different metal oxides reside at the other corners of the base triangle.

The diagrams yield much information for those familiar with their use, but the most important features from the phosphate chemist's point of view are compounds, melting temperatures, double salts, solid solutions, eutectics, phase transitions, miscibility gaps, and melting characteristics such as incongruent melting. Other factors such as compound stability, amorphous regions, phase compositions, heats of fusion or heats of transitions, and similar information can also be obtained when preparing the diagrams.

It is not enough to determine the phase diagram of a system and then to consider it a finished product. A phase diagram reveals nothing of molecular structure nor is it an analytical tool. For example, one may discover that a new compound $[YX(PO_3)_4]_n$ exists. The compound could be a double salt, a solid solution, a very long chain polyphosphate, or any of several ring metaphosphates ranging from trimetaphosphate to decametaphosphate or greater. Numerous techniques have been applied to discover the complexity of the compounds, but single-crystal X-ray analyses can be definitive.

5.6.1. Choice of Components

It was decided that in the Phosphate Fibers Project calcium, sodium, potassium, lithium, magnesium, and iron phosphates would be initial starting materials. Major emphasis would be placed on calcium because it is abundant, safe, and inexpensive. A logical first choice of a compound is a long-chain polyphosphate, $[Ca(PO_3)_2]_n$. It is reported to have an amphibole structure, but this is probably a loose use of the word amphibole. In silicate chemistry, amphiboles are double-chain structures with oxygen bridges between chains.[11] Pyroxenes are single chains of silicates and these are more nearly forms of calcium polyphosphates (often referred to as calcium metaphosphates) that are single chains of phosphate groups. Double chains are ultraphosphates and are highly susceptible to hydrolytic degradation.

5.6.2. CaO, P_2O_5 Systems

Figure 5.1 is a phase diagram of calcium phosphates from $Ca_2P_2O_7$ to $Ca(PO_3)_2$ and is truly a work of art. It is remarkable in its completeness and reliability.[12] From the viewpoints of phosphate fibers it is immediately recognized that a melting temperature of $[Ca(PO_3)_2]_n$ is 985 °C, but there are two forms of crystalline $[Ca(PO_3)_2]_n$, α and β. The transition temperature from α to β is about 970 °C, a value that is high enough to cause little or no problems provided a β phase is crystallized and not an α phase. This can be tricky, because it is not uncommon for a metastable phase to crystallize preferentially (Oswald's rule). If this happens the integrity of any fibers could be jeopardized when a transition occurs in a finished product.

Figure 5.2 is a presentation of Hill, Faust, and Reynolds's diagram containing two crystalline ultraphosphates, CaP_4O_{11} and $Ca_2P_6O_{17}$.[13] These are probably the first and only crystalline calcium ultraphosphates ever prepared. Many other crystalline ultraphosphates have been prepared and studied in great detail.[14] P_2O_5 itself is an ultraphosphate having both crystalline and amorphous forms. It will be noted that there are no corresponding sodium ultraphosphates in any of these phase diagrams. Instead, there is a region in which only sodium ultraphosphate glasses are obtained. This is very useful information when preparing $[Ca(PO_3)_2]_n$ fibers.

5.6.3. $[Ca(PO_3)_2]_n$ Crystals

Calcium polyphosphates are not innately long fibers and all attempts to grow long fibers from systems composed of pure calcium phosphates with M_2O/P_2O_5 ratios of unity were not very promising. Some very successful and important work is currently being done in Japan to grow well-formed short fibers of $[Ca(PO_3)_2]_n$ in neat systems. More attention will be given to this work in the final chapter of this book. The future of phosphate fibers probably resides in research laboratories in Japan.

After several unsuccessful attempts to grow long fibers, it became obvious that some different approach was needed. It was reasoned that if some method of causing

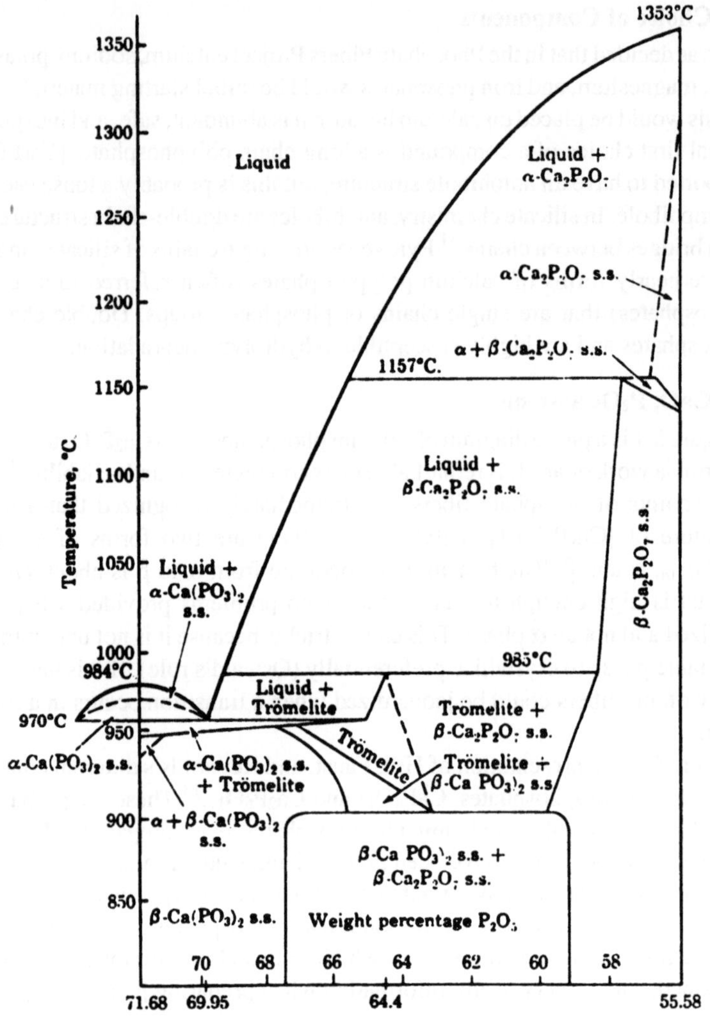

Figure 5.1. The phase diagram for the $CaO–P_2O_5$ system from $Ca_2P_2O_7$ to $Ca(PO_3)_2$.

crystals to grow dendritically could be found, it might be possible to grow these crystals in mixed salt systems. A question of which systems should be considered introduced more phase diagrams.

5.6.4. Na_2O, P_2O_5 Systems

Figure 5.3 is a phase diagram for sodium polyphosphates from tetrasodium pyrophosphate, $Na_4P_2O_7$, to sodium metaphosphate, $[NaPO_3]_n$.[13] At this point it is

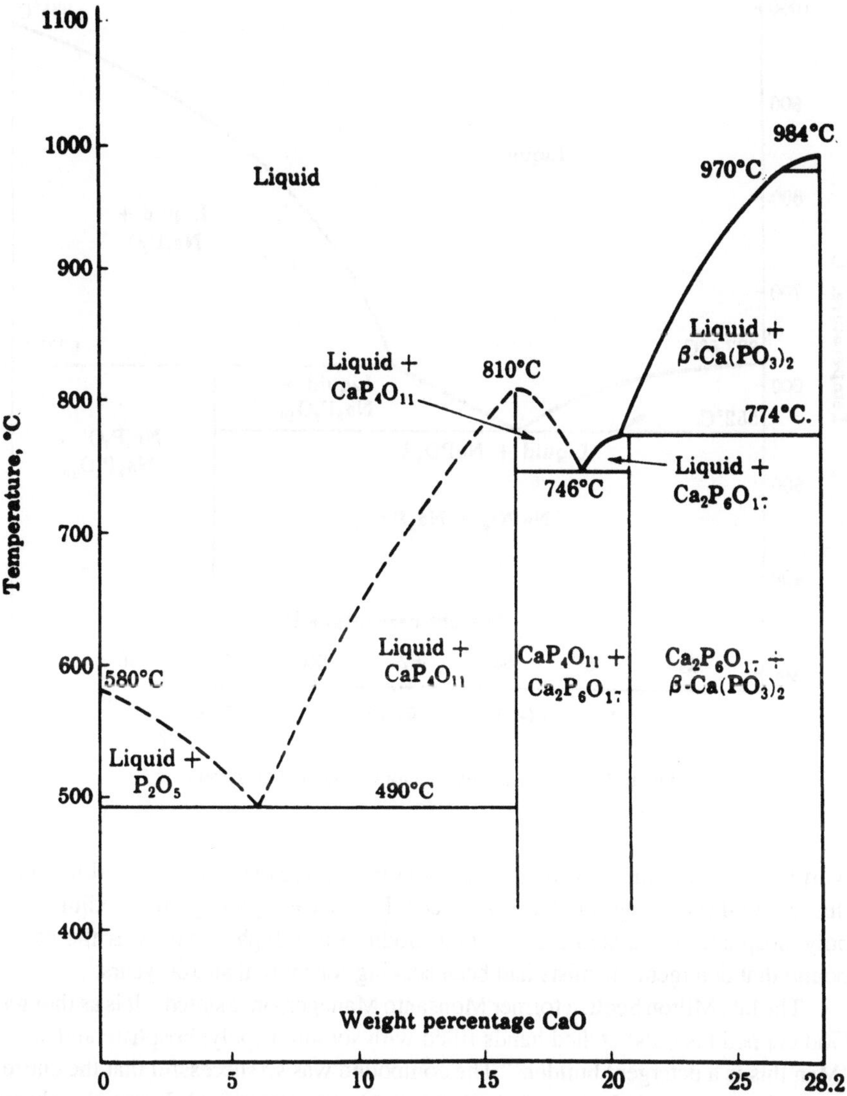

Figure 5.2. The phase diagram for the CaO–P_2O_5 system in the ultraphosphate region from $Ca(PO_3)_2$ to P_2O_5.

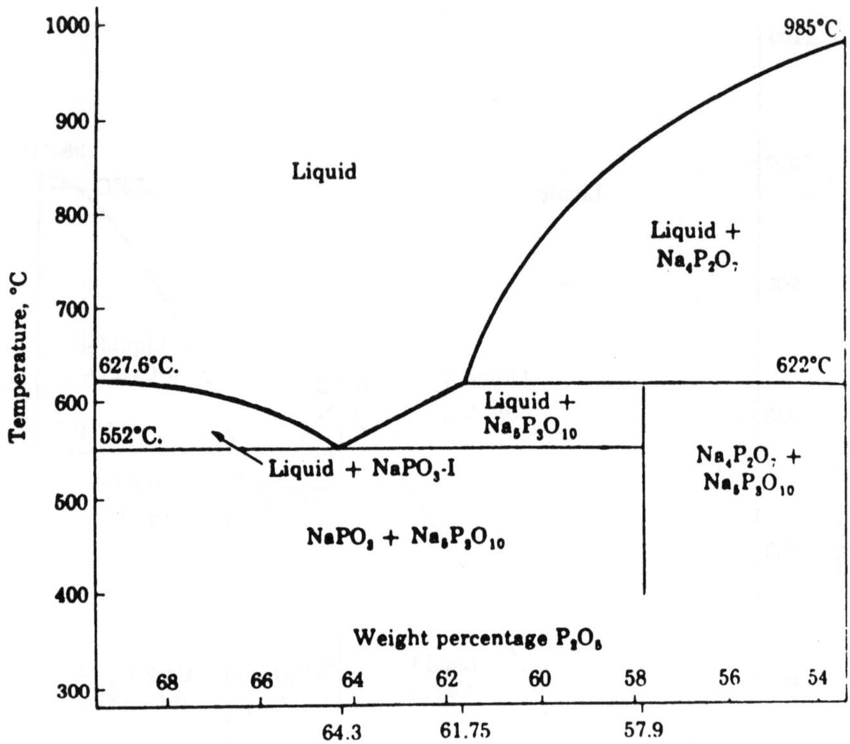

Figure 5.3. The phase diagram for the Na_2O–P_2O_5 system.

worthwhile mentioning this phase diagram for its value in a practical world. It was this Na_2O–P_2O_5 diagram that convinced Dr. Oscar Quimby that sodium tripolyphosphate was a stable compound. Sodium tripolyphosphate was the compound that detergent chemists had been seeking for more than 100 years.

The late Milton Scott, a former Monsanto Manager, once stated, "It is as though God cupped his outstretched hands filled with sodium tripolyphosphate and said, 'Man this is a detergent builder.'" The compound was so successful that the entire synthetic detergent industry was built upon this one compound. It was the phase diagram and the compound in the phase diagram that allowed the automatic home washing machine to become practical and inexpensive. Without sodium tripolyphosphate to lead the way it is highly probable that there would still be no successful home washing machines. A successful prototype must first be launched before inferior substitutes can be modeled to mimic the initial success, a practice not uncommon with "Me too!" mimics of successful consumer products.

Others had claimed that sodium tripolyphosphate existed as a compound before Partridge, Hicks, and Smith on the one hand and Morey and Ingerson on the other published phase diagrams demonstrating beyond any doubt that sodium tripolyphosphate existed.[15] The problems with former claims stem from the fact that just about every conceivable structure from monometa- to dimeta- to dodecapolyphosphate and above, and compounds in between, had been claimed by someone. Phosphate chemistry was in shambles. Van Wazer, more than anyone else, by publishing his classic papers began to bring some order to the systems. I worked under Dr. Van Wazer's direction and assisted him in giving structure to his ideas. Without his teachings I should never have been a phosphate chemist of any merit.

5.6.5. Na_2O, CaO, P_2O_5 Systems

More than 100 phosphate melt phase diagrams have been published.[15,16] There are probably several hundred more that have never been published. Only those that were directly related to phosphate fiber development and research will be reviewed. Several published phase diagrams were needed to initiate a formal study designed to prepare phosphate fibers. Perhaps the best starting place is to discuss a diagram base of a three-dimensional, three-component phase diagram, where components are Na_2O, CaO, and P_2O_5.

It makes little difference whether or not weight percent or mole percent is chosen as the units in a phase diagram, provided the units remain consistent. Most published diagrams have chosen weight percent as a unit, but we shall use mole percent on a base composition diagram. In diagrams presented here, units will be chosen in line with the authors who first published these works.

5.6.6. The Composition Basic Diagram

Figure 5.4 presents a base diagram for the three-dimensional diagram presented in Figure 5.8 This is a composition diagram, but it is used to extrapolate the behavior of $[Ca(PO_3)_2]_n$ when growing from melts of systems other than pure calcium phosphates. It can be seen that the $[Ca(PO_3)_2]_n$ field of dominance extends from $Ca_2P_2O_7$ to ultraphosphates, $Ca_2P_6O_{17}$, and CaP_4O_{11}. As one adds sodium as sodium phosphates to this system, the dominance extends in a more or less circular path. There is less tendency for $Ca_2P_6O_{17}$ and CaP_4O_{11} to crystallize, but this chemistry was never studied in detail, as no effort was made to determine the exact nature of the crystalline\amorphous soluble phases. It is possible that yet to be reported crystalline sodium calcium ultraphosphates exist under these conditions. A casual short probe of this system suggested that considerable skill would be required to obtain a reliable phase and composition diagram. This is particularly true because submerged phases similar to calcium penta-hexapolyphosphate, Tromelite, are likely to be encountered.[17–19]

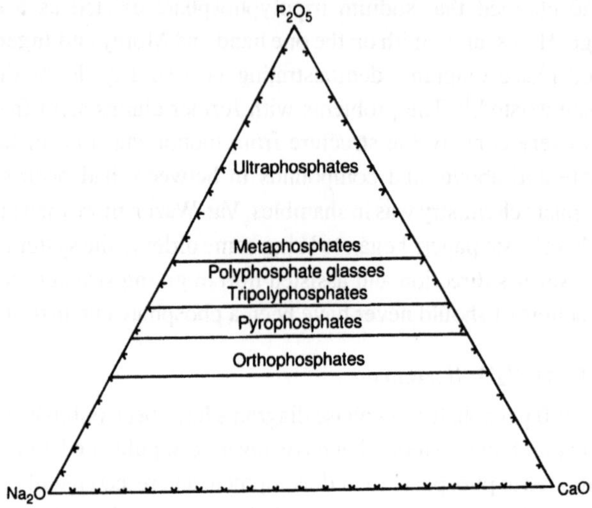

Figure 5.4. The base of a triangular Na_2O–CaO–P_2O_5 phase diagram.

5.6.7. The Three-Dimensional Diagram

In Figure 5.4 the corners of this diagram are chosen such that P_2O_5 is on top while mixtures Na_2O and CaO are represented by the bottom line. The first line above the base is all of the possible anhydrous sodium-calcium-orthophosphates. Above this line is the smallest sodium-calcium-*poly*phosphates, sodium-calcium-*pyro*phosphates. There are known crystalline tri-, penta-, and hexaphosphates in the sodium-calcium-polyphosphate system. The author could find no reliable evidence that a tetraphosphate of sodium, calcium, or mixtures of these two had been crystallized. A sodium tetraphosphate has been claimed from time to time, but has never been reliably demonstrated.

Above hexaphosphate compositions all other polyphosphates are amorphous, until the very long chain polyphosphate such as Kurrol's salts, Maddrell's salts, insoluble metaphosphate, IMP, and the like are reached. There is therefore an amorphous region of the phase diagram within the interior of which no known crystalline polyphosphates are known to exist. Perhaps these will be prepared as crystalline salts some time in the future.

Next, a line between poly- and ultraphosphates represents metaphosphate compositions. These are long-chain polyphosphates or ring compounds. Both types have been discovered and this line is very important for preparations of phosphate fibers. Above metaphosphate compositions are ultraphosphates. These are rich in P_2O_5 and *all* compositions in this area contain cross-linked phosphates, or triply linked POP linkages. The ultraphosphate region is important also in the preparation of crystalline phosphate fibers. It will be seen that not all compounds crystallized in this region must be ultraphosphates. Calcium polyphosphate fiber is not an ultraphosphate and it is crystallized in an ultraphosphate portion of its phase diagram.

Phase diagrams are seldom isolated stand-alone data resources. They are usually parts of larger systems that can often be daisy-chained, provided proper diagrams have been determined. If three or more diagrams have been published, they can often be reconstructed to become walls of a three-dimensional diagram from which good guesses can be made about a total system. Just a few measurements can yield much about an internal structure. Four diagrams were found that fitted our needs.

5.6.8. $Na_4P_2O_7$, $Ca_2P_2O_7$ Systems

If the ratio of M_2O to P_2O_5 is raised to two, the system becomes the $Na_4P_2O_7$–$Ca_2P_2O_7$, pyrophosphate system, shown in Figure 5.5.[20,21] Since pyrophosphates are the smallest members of polyphosphate systems, they are a logical choice for a back wall of this diagram under construction.

If two corresponding ends containing P_2O_5 are joined to form a triangle base, all three walls of a new diagram have been formed. Additionally, if the metaphosphate diagram in Figure 5.7 is inserted at the proper coordinates, a reasonably good picture of this entire system is obtained, but it is not a correct diagram for the system.

5.6.9. $[Ca(PO_3)_2]_n$, $[NaPO_3]_n$ Systems

Figures 5.6 and 5.7 are mixed diagrams of sodium and calcium polyphosphates, with a metaphosphate composition, ranging from $[Ca(PO_3)_2]_n$ to $[NaPO_3]_n$. Figure 5.7 could be considered a third wall of the diagram, but it is better to use Figure 5.5 for a back wall and use the metaphosphate diagram internally, to give added information about this system of three components. Components were chosen as CaO, Na_2O, and P_2O_5. The diagram that contains the two phosphate fibers, $[NaCa(PO_3)_3]_n$ and $[Ca(PO_3)_2]_n$, is the primary reason for writing this book.

There is a compound missing from Figure 5.6 that was later shown to exist and is shown in Figure 5.7. Sometimes, when seeds of a crystal have never existed before, systems can be very difficult to nucleate, but after crystals have been grown it is often difficult to avoid growing these crystals. This was probably the case with respect to Figure 5.6. More discussion will follow about this phase, because it will become a very important part of phosphate fiber systems.

Figure 5.8 is the completed phase diagram for the entire Na_2O–CaO–P_2O_5 phase system. It was assembled entirely from published data, except for the region of the diagram in Figure 5.7 where it was recognized that a compound was missing. This portion of the diagram was determined experimentally. It can be seen in Figure 5.8 that this diagram is completed and Figures 5.1, 5.2, 5.3, 5.4, 5.5, and 5.7 have

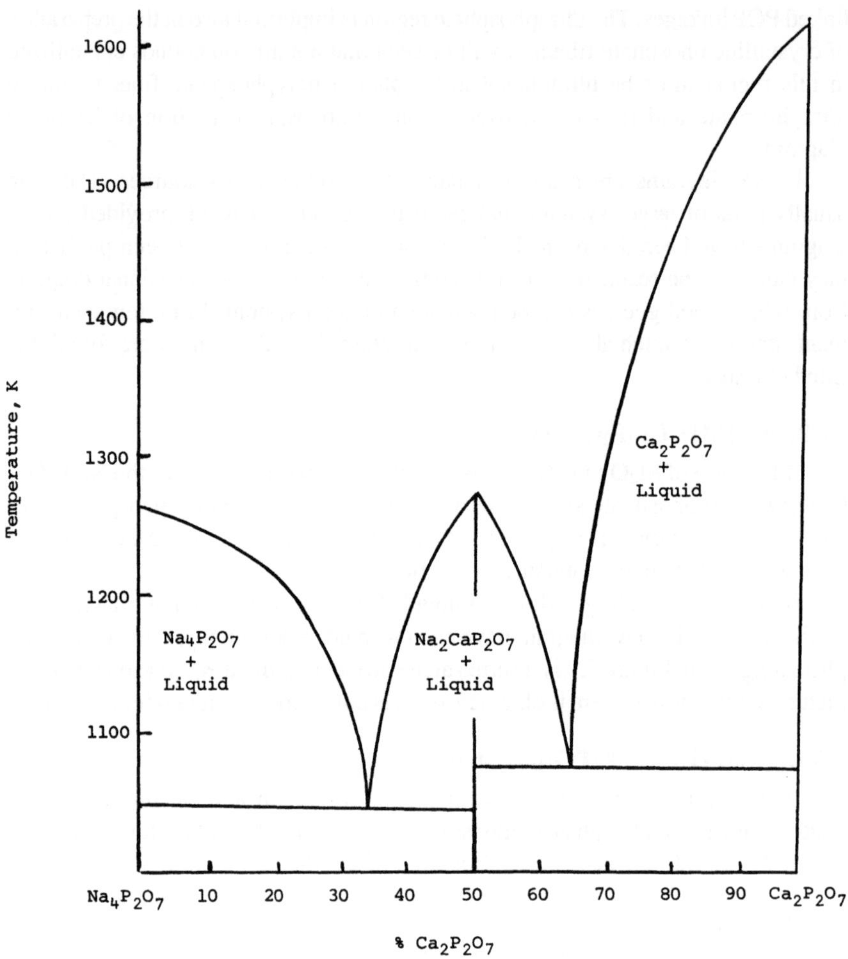

Figure 5.5. The phase diagram for the Na_2O–CaO–P_2O_5 system from $Na_4P_2O_7$ to $Ca_2P_2O_7$ based on the work of Berak and Znamierowska.[20,21] Note that this will become the back side of the phosphate fiber diagram (Figure 5.7).

been daisy-chained to yield a single diagram taken exclusively from superb published works. This approach can be continued indefinitely, since two of the components remain the same at the fields of juncture. As an example the K_2O–CaO–P_2O_5 could be joined to the diagram on the CaO–P_2O_5 side while the K_2O–Na_2O–P_2O_5 diagram could be attached to the Na_2O–P_2O_5 side of the diagram, all from published literature. Although these new diagrams would not be complete, they offer much scope for new experimentation.

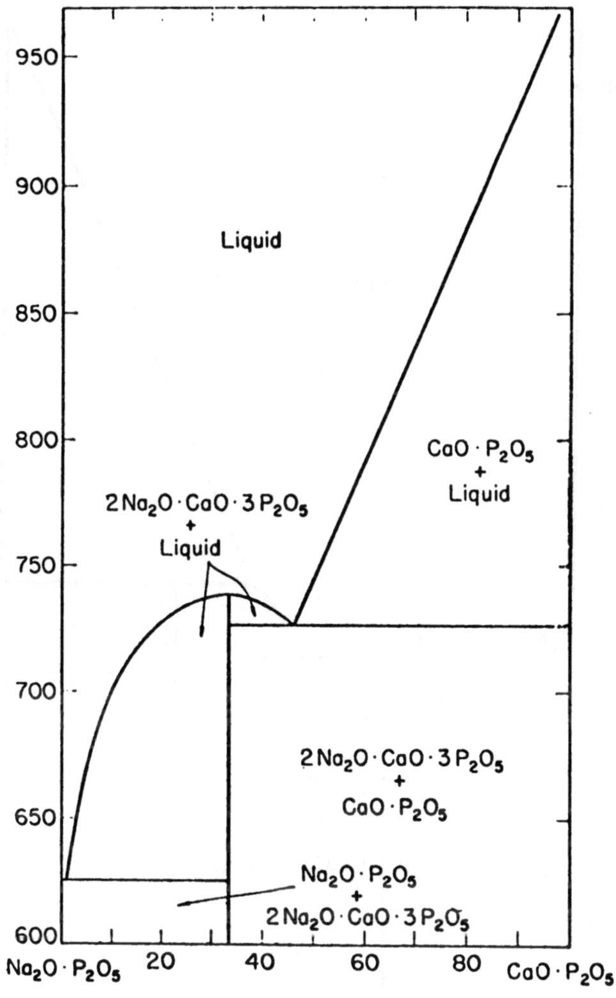

Figure 5.6. The phase diagram by G. W. Morey[7] of the Na_2O–CaO–P_2O_5 system from $Ca(PO_3)_2$ to $NaPO_3$. [Reprinted with permission from *J. Am. Chem. Soc.* **74**, 5783 (1952).]

When $[Ca(PO_3)_2]_n$ crystalline fibers had been made, a doping game began. It was expected that added substances should change the crystal morphology and rates of growth of crystals. There is small doubt that both influences occur and R. L. Hanson perfected this subject to a degree that some remarkable crystals were grown from fifty to one-hundred pound batches of melts. In many ways they behaved in a manner similar to long staple chrysotile.

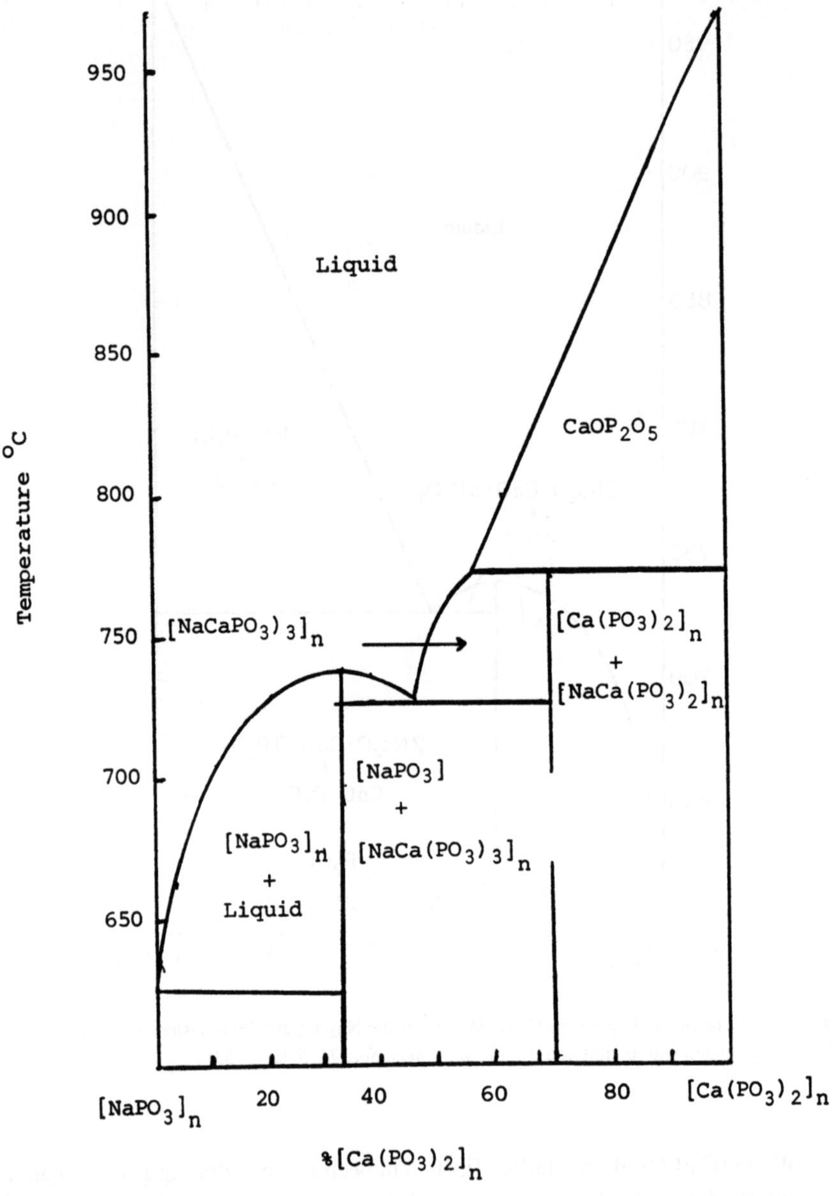

Figure 5.7. The phase diagram for the Na_2O–CaO–P_2O_5 system from $Ca(PO_3)_2$ to $NaPO_3$ with the correction for $[NaCa(PO_3)_3]_n$.

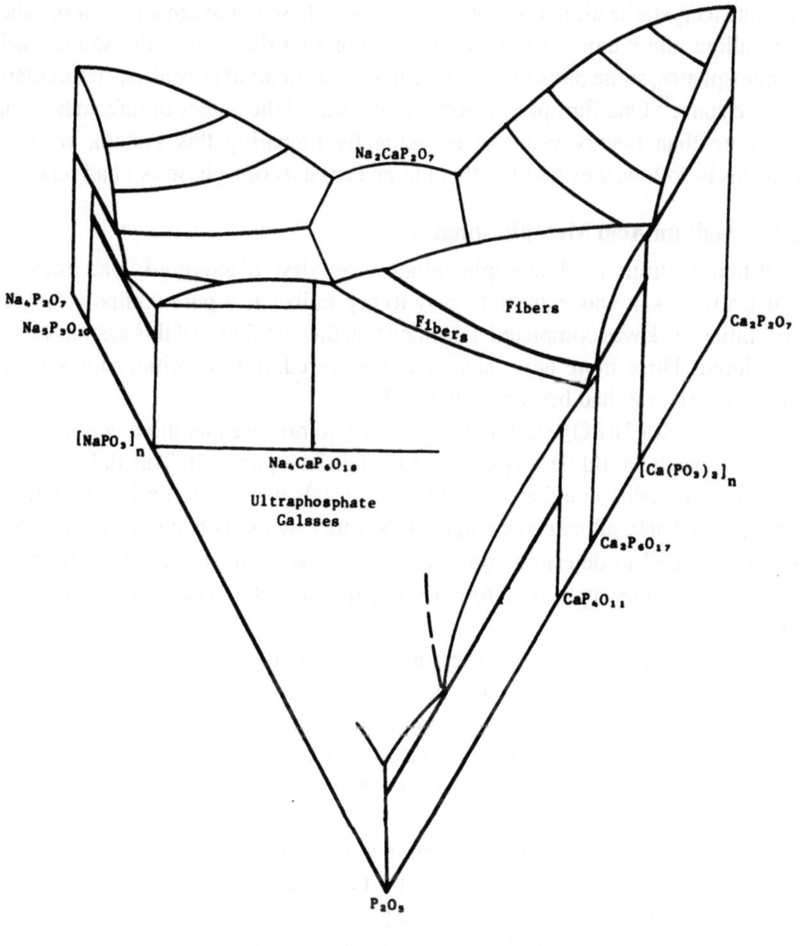

Figure 5.8. The phase diagram for the phosphate fiber phase system.

During the early days of growing calcium polyphosphate fibers it was recognized that fibers could be grown in a water-soluble melt, and that this had several advantages in not only growing crystals but also in extracting single fibers after compositions of crystallizing melts had reached some almost completed state. It was unnecessary that crystallization be totally completed. Any melt that did not crystallize to fibers merely became a part of a water-soluble phase between fibers, and would dissolve when fiber bundles were leached with hot water to free them.

With the current knowledge of melts it is impossible to predict whether or not a system will crystallize or remain amorphous when cooled. Before seeds are found to initiate a crystallization, it is not unusual for a phase to remain amorphous when a crystalline phase can exist. When the author first discovered the sodium acid metaphosphates, some parts of this system were difficult to crystallize, particularly in a pure state.[22] One European laboratory contacted the author to inform him that after more than twenty years of occasionally preparing this system, with my directions to follow, they had finally obtained crystals of a phase as I had reported.

5.6.10. Sodium Acid Metaphosphates

When sodium acid metaphosphate was first discovered, one form of $[Na_2H(PO_3)_3]_n$ salt was extremely easy to crystallize as a pure compound. After some initial work was completed, I decided that the remainder of this system should be explored. Three more new salts were discovered, but the initial supply of the first discovered salt had become exhausted.

The salt, $[Na_2H(PO_3)_3]_n$, had been so easy to prepare that there was no doubt that a new batch would be no problem to obtain as a pure salt. This did not prove to be the case. Salts on either side of these crystals were more easily crystallized than the initial salt, after seed crystals of the other two salts had been made. More salt was required to determine precisely its composition. It required almost two years of continuous repeated efforts to prepare the initial salt a second time in a pure condition.

The salt could not be prepared according to the instructions in my own notebook. I finally devised a new route to the crystals. Even then, this compound was obtained in very small amounts, compared to the several hundred gram samples that were obtained before this system became contaminated by other crystals of the same system. It was a hard lesson in the power of seed crystals in condensed phosphate systems.

Two of the new salts discovered in the sodium acid metaphosphate were fibrous with asbestiform crystals and both are respectable candidates for fiber applications. If the salt, $[Na_2H(PO_3)_3]_n$, was ever to be practical, a better way of crystallization was surely required. The salt $[Na_3H(PO_3)_4]_n$ was very easily made and had properties that should have been quality fibers. It had a very slow rate of solution and was similar to Maddrell's salt in many of its properties. The phase diagram for this system is presented in Figure 5.9. The diagram was particularly difficult to obtain because these salts were inclined to become unstable at temperatures approaching their melting temperatures.

Seeding is very important when growing condensed phosphates, and yet seeding is very poorly understood. Currently it is impossible to know, before a crystal is tried experimentally, whether or not it will induce crystallization in a melt. The first crystals of the sodium acid metaphosphate system were obtained while passing an electrical current through a melt of an amorphous salt. It was never known for certain if this was responsible for the crystallization, or was merely chance and crystallization would have occurred spontaneously without the electric

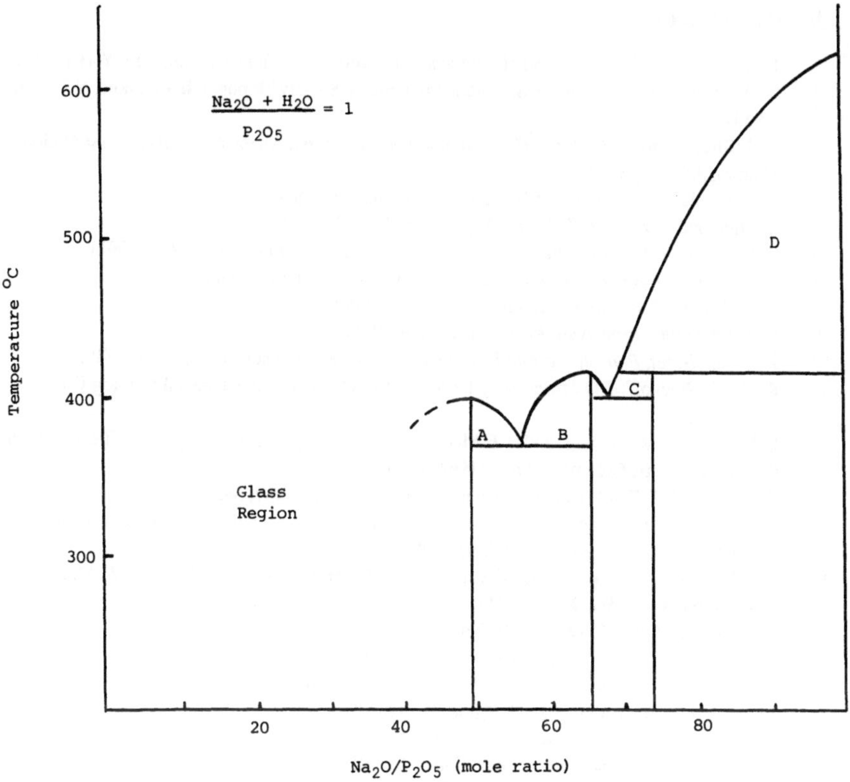

Figure 5.9. The phase diagram for the sodium acid metaphosphate system: area A contains $Na_2H_2(PO_3)_4$ plus liquid; area B contains $Na_2H(PO_3)_3$ plus liquid; area C contains $Na_3H(PO_3)_4$ plus liquid; area D contains $[NaPO_3]$ plus liquid.

current. Once seeds were present in the laboratory there was no method to repeat this experiment, because all these systems crystallized spontaneously, but slowly.

5.7. Conclusion

There are relatively few crystalline compounds that must be considered in the published phase chemistry of phosphates if only anions are considered. When all possible mixed systems of cations and anions are included, there are an enormous number that could give interesting results. The Durif school of phosphate chemistry in France has studied hundreds of systems in detail while the German school following Thilo has added great wealth to our knowledge of these mixed systems. N. N. Chudinova of Moscow State University has provided some exciting new knowledge to crystalline ultraphosphate chemistry.

5.8. References

1. J. E. Ricci, *The Phase Rule and Heterogeneous Equilibrium*, Van Nostrand, New York (1951).
2. G. W. Morey, *Properties of Glass*, American Chemical Society Monograph, Reinhold, New York (1954).
3. K. J. Shaver and J. G. Stites, *The Heteropolymeric System, $Na_2O–P_2O_5–SO_3$*, 131st National Meeting ACS (April 1957).
4. A. Schleede, W. Schmidt, and H. Kindt, *Z. Elektrochem.* **38**, 633 (1932).
5. E. Thilo and L. Kolditz, *Z. Anorg. Allg. Chem.* **278**, 122 (1955).
6. E. J. Griffith, *Cake Formation in Particulate Systems*, VCH Publ., New York (1991).
7. G. W. Morey, *J. Am. Chem. Soc.* **74**, 5783 (1952) and private communication.
8. G. W. Morey, *J. Washington Acad. Sci. J.* **13**, 326 (1923).
9. P. W. Bridgman, *Proc. Natl. Acad. Sci.* **21**, 109 (1935).
10. J. R. Van Wazer, *Phosphorus and Its Compounds*, p. 123, Interscience, New York (1958).
11. R. W. G. Wyckoff, *The Structure of Crystals*, pp. 322–349, The Chemical Catalog Co., New York (1931).
12. E. M. Levin, C. R. Robbins, and H. F. McMurdie, *Phase Diagrams For Ceramists*, The American Ceramics Society, Columbus, Ohio (1964–1975).
13. W. L. Hill, G. T. Faust, and D. S. Reynolds, *Am. J. Sci.* **242**, 547, 542 (1944).
14. J. Majling and F. Hanic, in *Topics in Phosphorus Chemistry*, Vol. 10 (M. Grayson and E. J. Griffith, eds.), p. 341, Wiley, New York (1980).
15. E. P. Partridge, V. Hicks, and G. V. Smith, *J. Am. Chem. Soc.* **63**, 454 (1941); G. W. Morey and E. Ingerson, *Am. J. Sci.* **242**, 4 (1944).
16. E. J. Griffith, *J. Am. Chem Soc.* **78**, 3867 (1956).
17. D. R. Gard, *Phosphorus Sulfur Silicon* **76**, 231 (1993).
18. J. R. Van Wazer and S. Ohashi, *J. Am. Chem. Soc.* **80**, 1010 (1958).
19. W. Wieker, A. R. Grimmer, and E. Thilo, *Z. Anorg. Allg. Chem.* **330**, 78 (1964).
20. J. Berak and T. Znamierowska, *Rocz. Chem.* **46**, 1697 (1972).
21. J. Berak and T. Znamierowska, *Rocz. Chem.* **46**, 1971 (1972).
22. E. J. Griffith, *J. Am. Chem. Soc.* **78**, 3867 (1956).

6

Laboratory Experiments

6.1. Preparations and Tests of Phosphate Fibers

In this chapter laboratory experiments will be discussed involving methods to prepare fibers with a polyphosphate structural framework. Since the anionic structure was selected for all systems under consideration, cationic types, ratios, thermal histories, seedings, and atmospheres dictate most of the variables that can be controlled.

There are many approaches to experimentation in a laboratory. One type that physical chemists engage in is the measurement of a known system to obtain new fundamental information about the system. These experiments are usually well planned and most of the equipment is in place before the first experiment is undertaken. High precision and attention to minute detail is essential if these works are to be worthy of completion. The kinetics of the hydrolytic degradation is such a study. The electrical conduction of solutions of condensed phosphates as a function of complexing cations and the determination of instability constants is another. The list could be almost endless and this science is usually conducted by highly trained personnel.

There is the type of research that is carried out to prepare new substances. It is research where something is done that has never been done before, to prepare a new substance or a known substance by a new approach. When a new material with new properties is being prepared for the very first time, it may require many approaches before the desired results are obtained. Organic and inorganic chemists engage in this type of research. They apply well known fundamentals to systems that were never before prepared in a laboratory or plant. This does not mean that these substances were never before prepared in the great laboratory of nature. The laboratory personnel that are successful are usually artists as well as scientists with extensive background in their science. New plastics, new paints, new drugs, new ceramics, and new textiles come from these experiments.

Formulation is a science and an art in a class by itself. The personnel who are successful in these research areas are very special. They usually have inborn talents that very often cannot be acquired by training. Perfume and flavor chemists and perfumers are of this class. They usually have a very good nose and a sixth sense for what will be appealing to many people. At times, in practicing formulation research tenacity is as important as knowledge. A formulation may contain so many different ingredients that science is incapable of dealing with all variables simultaneously. Under these conditions the formulator has but one choice: to roll up his sleeves and start mixing and observing. Shelf life, solubility, stability, light fastness, hygroscopicity, freezing points, odor, color, viscosity, taste, and function are but a few of the qualities often demanded by a customer. Most formulation work is done by personnel with a broad background and much experience, but who are willing to work by trial and error when necessary.

There is a type of research that is more invention than science. One has an idea of the results required to cause the project to become a success, but no one knows exactly what should be done nor exactly where to start. The early stages of the Phosphate Fibers Project was like this type of probing research, intended to identify a starting point in the system.

Methods of analyses are important in any type of research project, but they are critical in a project intentionally designed to trailblaze. Results of each completed experiment dictate the direction and complexity of subsequent experiments. This results in a maximum utility of resources while excluding interesting unproductive safaris into barren areas unrelated to final goals. This requires reliable means of judging results quickly, based upon meaningful analyses and testing. Without analytical chemistry and reliable testing no project can be a success.

As we continue to chide technical management they seldom seem to understand that in research one thing leads to another. It is a discovery that could not be anticipated when a project was initiated that yields great returns. One favorable result does not make a product, and the mere fact that something has been found that works does not mean that research should be terminated. Patent attorneys have recognized this for many years.

Something new, something worth doing in a research project is, "Not obvious to a person of ordinary skill in the art." This includes the inventor. Without these conditions, there is no invention and there are no patents. Today's research must dictate tomorrow's results. Business-type management would very much like for everyone to sit down on January 1, 199X, any year, and plan the exact experiments to be completed at noon on December 31, 199X. Yesterday I sat in a meeting with a Technology Director demanding a written program for the work to follow with dates for the exact time each phase would be completed. One may successfully direct a cotton mill by edicts of discovery, but this is a bedrock recipe for corporate waste of research funds and a long-term business plan for failure and bankruptcy. One seldom cooks a tasty cake by halving the cooking time and doubling the oven temperature of the recipe.

The "Business Team" approach to the discovery of new useful products is a myth. Invention and discovery is a lonely process usually conceived of and understood by some single highly motivated and dedicated person. This person usually has to work hard to get the other team members to understand the first successful experiments. A team is required to develop and implement the inventions and discoveries if the inventions are ever to be of use to humanity. A successful organization will recognize the differences between invention and implementation and will organize to optimize both processes. A team has never produced a great painting. A team has never written a great novel. A team has never written great music. A team has never fathered a genius child. Conversely, an individual has never built a giant skyscraper. An individual has never built a huge jet airliner. An individual has never constructed an interstate highway. Individual and team functions should be recognized and encouraged, despite the teachings of some recognized business schools whose teachers have never made an invention or a scientific discovery in their entire lives.

6.2. Inorganic Phosphate Chemistry

Inorganic phosphate chemistry is almost devoid of synthesis chemistry so highly developed in organic chemistry. In inorganic phosphate chemistry molecules are not assembled one part at a time. Instead, melt chemistry that is required to prepare new condensed phosphates is usually a crystallization process, controlled by phase chemistry and the dictates of the gods. With a phase diagram in hand, it is usually relatively easy to predict what types of molecules to expect, but even a phase diagram does not reveal structure and applications. Even when armed with phase diagrams, there are always likely to be some surprises both good and bad.

With experience and a few simplistic rules one can usually, but not always, predict the general types of molecules likely to be formed by crystallizing a melt of a composition never before prepared. For example, Griffith and Buxton discovered a crystalline twelve-membered ring, hexametaphosphate. For more than 100 years chemists had sought sodium hexametaphosphate as a true compound. When we discovered the true hexametaphosphate and a way to make it by crystallizing a melt, it came as a complete surprise.[1]

6.2.1. Sodium Hexametaphosphate

Thilo and his group in Germany discovered true crystalline hexametaphosphate almost simultaneously with Griffith and Buxton, but Thilo's approach was to extract meticulously this compound from Graham's salt where it was known to form as a trace component but had never been crystallized as a pure compound.[2] There is about as much sodium hexametaphosphate in the "Sodium Hexametaphosphate" glasses of commerce as there is dog in a hot dog. There is probably some of each, in each.

In lithium phosphate melts where lithium hexametaphosphate crystallized as a phase, melt composition should have yielded a linear hexapolyphosphate if everything had occurred as planned. We recognized that an attempt to prepare hexapolyphosphate was little more than a blind gamble. If lithium hexapolyphosphate had been a member of this phase family, all conditions were optimized for its formation, but it is not found as a stable alkali metal compound derived by melt crystallization. As noted below, calcium hexapolyphosphate can be crystallized from a melt, but usually with great difficulty.[3]

6.2.2. Hexapolyphosphate

Preoperative inorganic phosphate chemistry is not an exact science and rules remain vague, depending often as not on chance, if the systems have never been studied before. The calcium polyphosphate, Tromelite, is a mixture of calcium pentapolyphosphate and hexapolyphosphate.[4,5] I am unaware of a *crystallization technique* to prepare directly a soluble hexapolyphosphate as a pure crystalline phase. Hopefully, a melt system will be discovered and this can be accomplished. Hexapolyphosphate can be extracted from acidic mixtures and crystallized as the acridinium salt. All phosphates ever tested by the author can be crystallized as the acridine salt (see Figure 4.1). It was first noted by Quimby that very large cations, such as acridine and guanidine, assisted in crystallizations of condensed phosphates.[6]

6.2.3. Unstable Phosphates

It is understood that inorganic phosphate chemistry is anion dominated. Once an anion such as tetrameta-, tripoly-, hexameta-, etc. has been prepared as one of its salts, there is usually very little problem converting it to another cationic species without violating the integrity of an anion. In other words, it is usually not difficult to convert a sodium salt to a potassium salt even though the potassium salt is not a member of any potassium phosphate phases and cannot be crystallized directly from a melt. This is not always true, as will be discussed in Section 6.2.5.

6.2.4. Potassium Trimetaphosphate

Potassium trimetaphosphate does not form thermally by heating KH_2PO_4 to a melt and crystallizing the melt, but sodium trimetaphosphate can be made by heating NaH_2PO_4 to a melt and crystallizing it. Potassium trimetaphosphate can be prepared by ion exchange techniques, applied in aqueous media, utilizing a trimetaphosphate anion derived from sodium trimetaphosphate.

Potassium trimetaphosphate can be prepared directly by heating KH_2PO_4 with urea acting as a dehydrating agent at temperatures near 200 °C.[7]

$$3KH_2PO_4 + 3NH_2CONH_2 \Rightarrow K_3P_3O_9 + 3CO_2 + 6NH_3$$

In a more basic melt it is known that urea is capable of converting some M_2O of a system to M_2CO_3, provided reaction temperatures are low enough that M_2CO_3 does not react with more condensed phosphate. An example is

$$n\ K_4P_2O_7 + n\ NH_2CONH_2 \Rightarrow 2[KPO_3]_n + n\ K_2CO_3$$

This reaction is not complete in experiments performed by the author, but since K_2O is definitely a chain breaker or a ring opener, it can be seen why potassium trimetaphosphate can be formed thermally, at lower temperatures, where it would not normally be expected to form. Formation of K_2CO_3 removes K_2O from the reaction zone. This reaction is similar to a well-known reaction of NH_4Cl with basic phosphates to make MCl and a more condensed phosphate. This lower-temperature chemistry is worthy of more study and should yield many surprising ways to make compounds that are not phase diagram entities, as they are prepared from melts.

6.2.5. Sodium Tetraphosphate Does not Crystallize

It is not always possible to prepare crystalline salts of an anion merely because an anion can be prepared. Lead tetrapolyphosphate is relatively easily obtained and lead tetrapolyphosphate can be easily converted to crystalline ammonium tetrapolyphosphate by treating this lead salt with ammonium sulfide.[8,9] As a result of this ease of preparing crystalline potassium trimetaphosphate from sodium trimetaphosphate, one might expect that it is an easy chore to prepare crystalline sodium tetrapolyphosphate from crystalline ammonium tetrapolyphosphate. To date, no one has ever prepared crystalline sodium tetrapolyphosphate, though several have claimed this success. Hopefully, someone will eventually master this system, but it will require a better understanding of these systems than currently exists.

When a phase diagram has been prepared one's chances of controlling products of a reaction are improved, but even this is not absolute because one can still miss a compound, as Morey missed the phosphate fiber, $[NaCa(PO_3)_3]_n$, in his diagram of $NaPO_3$–$Ca(PO_3)_2$ systems.[10] This is in no way intended to fault Morey. The seeds of these crystals did not exist and they did not happen to form under his initial conditions. It is merely more of an indication that even the most experienced investigators are always at the mercy of chance when dealing with new inorganic phosphate chemistry. Durif's group in France discovered the missing compound in the diagram during the same time frame that we discovered it in our laboratory.[11]

6.2.6. Amorphous Phosphate Fibers

Much work has been expended on unsuccessful attempts to prepare useful spun fibers of phosphate glasses similar to silicate fiber glasses. No problems exist in obtaining spun fibers. It is relatively easily accomplished. The primary problem that has been encountered with glassy phosphate fibers is their tendency to crystallize and become weak, or their tendency to become hygroscopic and weaken from hydrolytic molecular degradation. Some beautiful fibers have been spun, but until now they have had very little utility in material science when long fiber life is required.

Some spun fibers are reported to be very useful in bone implants and similar applications, where it is desirable that a fiber degrade and be absorbed in a relatively short time. Here, these fibers would lend green strength to a composite that is ultimately replaced by normal bone.[12] Some more recent systems are also showing some promise as high-temperature spun amorphous phosphate fibers. This utility may eventually be possible.

6.3. Safety, Safety, Safety, Always Safety

The primary useful property of phosphates that perhaps nothing else approaches, is the safety record of phosphates in manufacturing plants and households around the world. It was with this fact in mind that two candidates for new mineral fibers were chosen. There were at least eight other candidates that could have been selected. As noted, there are two general molecular forms of serpentine minerals that are referred to as "asbestos." These are the chrysotile and the amphibole types. These two forms of "asbestos" are about as similar, chemically and physically, as sodium chloride and sucrose; both are white crystalline solids.

Phosphate chemistry and silicate chemistry are much alike, but they are also very dissimilar. It was the expressed purpose of this work to prepare crystals of phosphates with properties similar to silicates referred to as "asbestos," by substituting phosphorus atoms for silicon atoms in molecular structures like those found in serpentine minerals. It was hoped that many useful properties of silicates could be gained while retaining the long safety record of condensed phosphates.

6.4. Calcium Polyphosphate, $[Ca(PO_3)_2]_n$, Preparation

Since the safety history of calcium phosphates is well known, it was chosen as a material of first choice. Raw materials were lime, CaO, and phosphoric acid, H_3PO_4, both of which have been used in foods throughout the civilized world for hundreds of years. It was decided to attempt the following reaction for a start:

$$n\mathrm{CaO} + 2n\mathrm{H_3PO_4} \Rightarrow [\mathrm{Ca(PO_3)_2}]_n + 2n\mathrm{H_2O}$$

To make life simpler, what one actually does in the laboratory is to heat monocalcium orthophosphate to drive out water contained in the melt and then slowly cool the system to crystallize $[Ca(PO_3)_2]_n$.

$$nCa(H_2PO_4)_2 \Rightarrow [Ca(PO_3)_2]_n + 2nH_2O$$

Care should be exercised in using monocalcium orthophosphate sold as Reagent Grade chemicals. $Ca(H_2PO_4)_2$ is unstable, being inclined to disproportionate into $CaHPO_4$ and H_3PO_4, because it is usually sold with an excess of Ca^{+2} or an excess H_3PO_4 in the crystals, to stabilize the salt. It is therefore necessary to analyze each batch of raw materials before it is used, if a proper Ca–P_2O_5 ratio is to be maintained. Exact ratios can be critical in some parts of a phase diagram.

6.4.1. The First Try Always Fails

A first try was nothing more than heating about 200 g of monocalcium orthophosphate, contained in a platinum dish, to a melt near 1000 °C. This system was then slowly cooled and crystals formed, as expected. These crystals were removed from the dish as a dense cake about 2 cm thick. The cake was crushed and milled in various ways to determine if these crystals could in any way be considered fibers.

Upon analyzing our experimental results, with some imagination, perhaps a few particles viewed under a microscope might be judged to be fibers. This first experiment could only be considered a failure, but several suggestions for subsequent experiments were revealed by the results. It was obvious that the crystals were grown as too concentrated a mass to be easily milled to single fibers. Also, these crystals were inclined to grow into each other as twins, harsh blades, or other random deformities. It was at once obvious that fibers must be grown in a more dilute melt system, but in what?

After a few more failures at attempting to grow $[Ca(PO_3)_2]_n$ crystals in inert salt systems such as Na_2SO_4, NaCl, and other salts showed little or no promise, the idea of growing fibers in a sodium ultraphosphate system was born. Sodium ultraphosphate melts are difficult or impossible to crystallize. They are very unstable to hydrolytic degradation and are easily soluble in warm water. Moreover, while considering combined phase diagrams of sodium and calcium phosphate systems it was obvious that there was an extensive region of these diagrams where $[Ca(PO_3)_2]_n$ should be the only crystalline phase, provided crystallization did not proceed to absolute completion. Theory predicts that a eutectic composition between crystalline phases is the last liquid to crystallize on cooling a melt system (see Figure 5.1).

6.4.2. Ultraphosphates Save the Project

At first encounter, the concept of growing $[Ca(PO_3)_2]_n$ fibers in an ultraphosphate melt seems a minor adjustment in conditions. There is a gigantic change in nearly all properties of this system with but slight changes in composition in moving from a melt of *poly*phosphate composition to a melt of *ultra*phosphate composition. These changes are most apparent in changes in the aqueous solution properties of water-soluble salts such as potassium Kurrol's salt grown from melts, in which the K_2O–P_2O_5 ratio is adjusted from 1.01 to 0.99. In the first case, solutions are typical of a very long chain polyphosphate. In the second case, the same concentration of potassium Kurrol's salt is a very viscous stringy solution with many fluid properties vastly exaggerated.[13]

The consideration of moving from a polyphosphate melt to an ultraphosphate is critical enough in understanding this system to include another example of the great change that can occur in crossing this composition boundary. It is not necessarily a phase boundary unless the systems crystallize. It has long been recognized that ultraphosphates are usually much less corrosive to metals in anhydrous systems than are polyphosphates. This is particularly true of the acids. The reason is straightforward. Polyphosphates are exceptionally good sequestering agents for many cations and will dissolve oxide coatings from metals. Ultraphosphates and metaphosphates (rings), on the other hand, are not particularly good sequestering agents. It is noted, of course, that to arrive in an ultraphosphate region of a phase system from a polyphosphate system, compositions pass through the metaphosphate system.

Inorganic phosphate chemists have normally treated their systems in much the same conventions in drawing structures and placing cations as organic chemists have done. In melts and glasses this is probably justified, but in a solid crystalline state it is well known that placement of cationic functions in structures is an oversimplification. Nevertheless, we shall follow the conventions for convenience, while recognizing that placement of anions and cations is often not a good representation.

Polyphosphate has been defined in this work as any linear condensed phosphate in which a phosphate, but not necessarily the system containing phosphates, exhibits a condition $1 \leq M_2O/P_2O_5 \leq 2$, where M is any single or mixed metal with a total equivalency of unity. This definition is required to differentiate a total composition of a system from polyphosphates crystallizing from a melt.

6.5. Preparations of Calcium Polyphosphate Fibers

It should be noted that calcium polyphosphates are often called metaphosphates in the literature and patents, if they have M_2O–P_2O_5 ratios near unity. Throughout this work, the word metaphosphate will be reserved for ring phosphates while chain phosphates will be named polyphosphates. In most modern usage the prefix "poly-" has unfortunately been dropped from a name. Calcium tripolyphosphate, $Ca_5(P_3O_{10})_2$, is named calcium triphosphate, which can easily be confused with tricalcium phosphate, $Ca_3(PO_4)_2$, an orthophosphate, by those who only occasionally engage in phosphate chemistry. In this work polyphosphates shall be named polyphosphates, unless reference is made to work where polyphosphates are named without the prefix. Here, they will be named as in the cited literature.

Cookbook examples of methods of preparing calcium polyphosphate fibers are given in patents granted for these compositions of matter. For those interested in growing their own crystals, Example 1 of United States Patent No. 4,360,625 of November 23, 1982 will be repeated here.

> "To 72.35 grams of calcium carbonate were added 166.25 grams of 85% phosphoric acid, 20.67 grams of monosodium dihydrogen phosphate monohydrate, and 8.27 grams of 85% phosphoric acid. The above ingredients were thoroughly mixed and placed in a 500 cc. platinum dish and heated to about 200°C. for approximately 2 hours to drive off residual moisture. Then the temperature was increased to 1050°C. for 1 hour. Thereafter, the temperature was decreased to 800°C. and a small crystal of potassium Kurrol's salt was placed in the melt near the edge of the platinum dish. Crystallization was complete after about 15 minutes. The platinum dish was allowed to cool to about 100°C. and the contents were transferred to a 4-liter containing about 2 liters of boiling water. Approximately 50 percent of the fibers separated from the mass in about 4 hours. The fibers were characterized as being smooth, cylindrical crystals having an average diameter of about 20 microns and being about 6.5 centimeters average length."

Detailed analyses, including X-ray powder patterns, are given in this patent. Several batches were analyzed and all were found to be almost exactly the same as ASTM spectra for β $Ca(PO_3)_2$, calcium (polyphosphate) metaphosphate.

Perhaps the results of growing fibers in an acidic melt can best be demonstrated with an alkali metal phosphate where the additional complexities of multicharged cationic functions can be avoided. The behavior of potassium Kurrol's salt has been treated in detail, but a treatment for systems containing mixed metal ions is unnecessary for this purpose.[14] In preparing $[Ca(PO_3)_2]_n$ in sodium ultraphosphate melts it will be assumed that all Na_2O is retained in the amorphous phase while most, but not all, Ca will report to the fiber phase. There is some quantity of $[Ca(PO_3)_2]_n$ solubilized in a sodium-rich ultraphosphate phase. This solubility is crucial in preventing Ca from randomly behaving as chain terminators on crystal surfaces. However, it is relatively simple to calculate an approximate percent of phosphate that will become amorphous or crystalline merely by assuming that solubilization of $[Ca(PO_3)_2]_n$ in a sodium ultraphosphate phase is very small.

Not only composition and thermodynamics can influence growth of polyphosphate chains, but thermal history of phosphates can often lead to metastable equilibria. Slow crystallization rates usually favor the formation of longer-chain polyphosphates, if a crystallizing salt is a thermodynamically stable crystal under the growth conditions. During crystallization, impurities (chain growth terminators) are driven to the surface of a crystal if neither solid solutions nor double salts are formed between impurities and crystallizing polyphosphate. Slower growth rates provide longer times for impurities to migrate to a crystal surface.

In systems where a crystallizing salt is not the thermodynamically stable salt, a system can be cooled too slowly to obtain maximum chain lengths of the crystallizing polyphosphate. Sodium Kurrol's salt is unstable with respect to sodium cyclic triphosphate under all conditions, at atmospheric pressure. If a crystallizing melt forming sodium Kurrol's salt is cooled too slowly, it converts sodium Kurrol's salt to sodium cyclic triphosphate. Proper seeding and rates of crystallization are important in obtaining high yields of sodium Kurrol's salt where chain lengths may be 5,000 to 10,000, or more.[15]

6.5.1. Chain Growth Terminators

Metal oxides (and water) are chain growth terminators of long-chain polyphosphates. Long-chain polyphosphate chains are very sensitive to chain terminators. The difference between two polyphosphate anions of chain length 1000 and one anion of chain length 2000 is only one M_2O group per two thousand PO_3 groups. Also, two 1000 chain length anions have the same average chain length as one 1998 anion and one pyrophosphate group.

Long-chain polyphosphates crystallize from melts to form straight molecule–ions aligned parallel to the longest axis of a crystal (Maddrell's salt) or as long helical chains (Kurrol's salt), aligned parallel to the longest axis of a crystal. In neither case are these molecules twisted or folded in crystal. This is important when considering how long-chain molecule–ions grow from melts.[16]

6.5.2. Phosphate Glass Chain Lengths

Attempts to obtain amorphous polyphosphates with average chain lengths of 1000 directly by quenching melts have not been successful. Average chain lengths of amorphous polyphosphates of near 400 have been the maximum obtained directly from melts. Sodium Kurrol's salt of average chain lengths of several thousand have been melted to make quenched glass. Chain lengths are no longer than those obtained by melting high-purity sodium (cyclic triphosphate) trimetaphosphate. Indirect methods of obtaining very long chain length amorphous polyphosphates have been devised.[17]

If calcium phosphate fibers grown from ultraphosphate melts are slurried with a solution of sodium ethylenediamine tartaric acid, EDTA, calcium phosphates crystals can be converted to an amorphous sodium phosphate that dissolves slowly, yielding very long chains in solution. Methods to determine the average chain lengths of these ultralong polyphosphates had not been approached before the project was terminated. Several approaches ranging from intrinsic viscosity, gel chromatography, centrifugation, and light scattering were planned. There is no doubt that the chain length is very long, but how long is still a matter of conjecture.

6.5.3. Crystal Habit Modifiers

Once the ratio of sodium ultraphosphate to calcium polyphosphate had been more or less standardized, the game of adding crystal habit modifiers was undertaken in an attempt to determine the influence of modifiers. No reliable generalizations could ever be stated as to how foreign ions influenced these systems, but there was no doubt that properties of melts and crystals could be changed. One of the major problems in this approach is finding additives that did not contribute to the toxicity of these systems. Elements such as nickel, copper, cadmium, cobalt, chromium, lead, etc. were strictly forbidden, as were many anionic functions that were known to degrade polyphosphate or were toxic themselves.

If the objective is to prepare the safest mineral fiber that it is theoretically possible to manufacture, no trace of a toxic substance can be tolerated! There was never any doubt in the author's mind that only food grade raw material could be used in this process, although the question of using purified wet phosphoric acid, derived from fertilizers, was raised from time to time. In my judgement, the best commercially available purified wet phosphoric acid is inferior to furnace acid and, although it probably could be used, it would be done with some loss of a safety margin that has been paramount throughout the history of this project. Safety is the hallmark of phosphate fibers. When safety is not an issue, some other fibers have superior properties.

In the author's attempts to find substances that would modify properties of calcium polyphosphate fibers, additions of magnesium oxide or magnesium phosphates seemed to have the most favorable influence, but results were erratic. It seemed that sometimes there were vast improvements and, at other times, no influence could be seen. Answers to this behavior may have been similar to Mr. Ngo's discovery that two crystallizations were required to prepared outstanding fibers of $[NaMg(PO_3)_3]_n$. These melts were first crystallized, remelted, and then crystallized again to yield some of the more beautiful fibers made in this work. Unfortunately, the doping of calcium phosphate remains a mystery. This project was terminated before it was completely understood.

6.6. Sodium Calcium Phosphate Fibers

The first preparations of calcium polyphosphate fibers were orchestrated following a preplanned approach that was altered as results demanded. Conversely, sodium calcium polyphosphate fibers were discovered. Rather than following a calculated scientific logic we were exploring. We were prospecting a phase diagram around calcium metaphosphate compositions. This discovery was blind luck and it was not obvious that these crystals contained fibers until they were milled for analyses.

Once the fibrous nature of this system was obvious, work was then initiated to improve the properties of these fibers. The procedure for making laboratory samples is outlined below as described in the patent.[18]

"To 76.29 grams of calcium carbonate were added 150 milliliters of water, 148.15 grams monosodium dihydrogen phosphate monohydrate and 232.13 grams of 85 percent phosphoric acid. The mole ratio of ingredients was 19.0 percent Na_2O, 28 percent CaO and 53.0

percent P_2O_5 The above ingredients were thoroughly mixed in a container, a 1000 cc glass beaker that was placed on a hot plate to evaporate excess moisture. Then a 500 cc platinum dish containing the above dish was placed in a furnace at 200°C for 80 minutes to drive off excess moisture. Then the temperature was increased to 800°C and maintained at that temperature for 1 hour. Thereafter the temperature was decreased to 600°C and maintained at that temperature for 1 hour. The platinum dish and the contents were allowed to cool to about 100°C and the contents were transferred to a 4-liter beaker containing about 2 liters of boiling water. After about 3 hours, hexagonal shaped blocks were removed by filtration. The blocks were washed with water and methanol, and ground in a mortar and pestle to fiberize the blocks into asbestiform crystals."

When laboratory work demonstrated that it was possible to prepare $[NaCa(PO_3)_3]_n$ consistently, with uniform properties, work was turned to other venues. Platinum dishes were reasonably good containers in which to prepare laboratory samples, but it was obvious that an industrial process could never withstand the expense of platinum-plated equipment.

As can be seen above in the description of how the fibers were made, they had been heated to a melt in much the same manner that calcium polyphosphate acicular fibers had been grown. No method was ever envisioned that would allow one to grow long calcium polyphosphate fibers without heating systems to temperatures that would completely melt a phosphate charge. Sodium calcium polyphosphate fibers were not grown as fibers, but as small blocky crystals that were milled to fibers. To grow blocky crystals offers much more flexibility than attempting to grow bundles of individual fibers having acceptable morphologies.

6.6.1. Calciners to Avoid Melts

Industrially, many phosphate salts are manufactured in a calciner. Industrial plant calciners range from rotating tubes 12 in. in diameter and 6 ft. long to monsters 12 ft. in diameter and 200 ft. long, very similar to a portland cement kiln. Some calciners are externally heated. Some are countercurrently heated, where a flame impinges from one end of a tube and feed enters the other end. Then there are co-current calciners, in which feed and heat enter the same end of the tube. Other variations occur in which some side stream product is recycled, or a calciner is a rotating pot and not a tube at all. There are many configurations and options in choosing a calciner.

A question arose, "Is it possible to grow crystals of $[NaCa(PO_3)_3]_n$ in a calciner?" If a calciner were first loaded with quality crystals of sodium calcium polyphosphate and heated to temperatures high enough to drive water from the system, a phosphate melt should form on the surfaces of these crystals. If this were the case, it should be possible to crystallize this melt to bed crystal surfaces. This is similar to the way synthetic gems have been grown, by growing boules of stone while feeding feed stock on top of a seed crystal held at temperatures slightly above fusion.

An advantage of utilizing a bed of sodium calcium polyphosphate is derived from an enormous, very hot surface area derived from millions of small crystals rotating in a heated calciner tube. It is possible to feed solution on to a bed at a fairly rapid rate without forming enough melt on crystal surfaces to either corrode calciners or cause sticking and lumping of calciner burdens. There was every reason to believe that this concept had a chance to work. If a calciner approach would work, the unwelcome prospects of designing around high-temperature melts could be abandoned forever.

A small calciner was made of stainless steel. The calciner was a tube about 6 in. in diameter and 18 in. long. It was rotated from a shaft that had been welded to the back side of a can. This tube was heated externally by a bank of laboratory bunsen burners or by mounting the tube in a muffle furnace with its shaft protruding behind the furnace. Depending on requirements, either approach worked well, but a furnace gave better temperature control. A charge was added to the tube and was heated to the desired temperature. When this chosen temperature was reached, an aqueous solution of sodium and calcium orthophosphates was sprayed onto the hot bed. After some adjustments, it was demonstrated that a rate could be found allowing crystals to grow at a reasonable rate. The product looked great, but unfortunately it was green, indicating that there was corrosion of the stainless steel tube.

6.6.2. Corrosion Problems in Calciners

Logic dictates that when corrosion occurs in a calciner, one should lower the temperature to determine if a satisfactory product can be obtained. This was done and corrosion became much worse. It was soon learned that the bed was much less corrosive to steel at higher temperatures where melt crystallized to crystals more quickly. When optimum temperatures and feed rates were determined, high-quality crystals could be made without causing harm to the calciner. This process was scaled up to pilot plant size and will be discussed in more detail in Chapter 7.

6.7. Maddrell's Salt

Along the way we have mentioned other long-chain sodium polyphosphates. In science there is one known as Maddrell's salt. In industry this salt is called IMP, which stands for insoluble sodium metaphosphate. Like so many salts in industry it is a complete misnomer. It is not insoluble. It is a polyphosphate with a metaphosphate composition, just as the other phosphate fibers are polyphosphates with metaphosphate compositions. It is sodium and it is an extremely interesting substance, because it is completely harmless and is ever so slowly soluble.

6.7.1. Toothpaste and Phosphate Fibers

Depending on how old you are, you may or may not remember Ipana toothpaste. It was an excellent toothpaste and was in widespread use before people decided that they would prefer to brush their teeth with SiO_2 because one can read the newspaper through a clear gel paste. After having worked on dentifrices for several years, I never understood why we wanted a clear gel for a toothpaste. It contributed nothing to oral hygiene. Other places you encounter sodium IMP phosphate fibers is in processed cheese. It is sold under the trade name Emulsiphos® and is an excellent nutrient.

6.7.2. Cloth That Cannot Burn

There is an evergreen problem with the interior of aircraft, hotel mattress covers, children's sleepwear, and the like because they are flammable. It is obvious that a cloth that could not burn and was safe enough to eat and brush our teeth with should be ideal *if crystalline fibers long enough to spin into yarn could be grown.* I was already calling the new–old fibers, "Kotton." It would be easy to spin amorphous fibers into fabric, but until now their long-term utility must be questioned. There is hope that this problem will be solved and no known reason why it cannot be solved.

To spin a crystalline short-staple fiber into yarn, fibers should be about one inch long. They could be used as a shorter staple, but they would be better if they were at least one inch in length. To have fibers that are one inch long requires that crystals be grown that are at least one inch long. Like sodium calcium polyphosphate fibers, Maddrell's salt grows as blocky single crystals that must be milled to form fibers. Milling and crushing become an important issue.

When corporate management publicly announced that they were abandoning the Phosphate Fibers Project, without mentioning it to the inventor before they made the announcement, I was growing crystalline sodium polyphosphate fibers of more than one-half inch. They were almost long enough to try the first spinning experiments. It was obvious that if as little as 5% long staple cotton fibers were added to the phosphate fibers, thread could be spun that would not burn. I did not like the cotton crutch. I wanted cloth that behaved like cotton, but was pure phosphate.

It is obvious that sodium calcium polyphosphates could be used as well as Maddrell's salt to make cloth, but sodium calcium polyphosphate had not been used in foods as crystals, despite the fact that they had been used in drinking water for more than fifty years. The difference would surely have been cosmetic, but it was felt that critics should have been completely disarmed when they recognized that sodium polyphosphate fibers had been a part of their diet for most of their lives.

This project has nothing to do with "asbestos" and should be reconsidered by a knowledgeable group, more interested in humanity's wellbeing. There are surely

Table 6.1. The Physical Properties of Calcium Polyphosphate Fibers Compared to Chrysotile Fibers

Property	Crysotile	$[Ca(PO_3)_2]_n$	$[NaCa(PO_3)_3]_n$
Crystal Class	Monoclinic	Monoclinic	Triclinic
Density (g/cc)	2.5	2.872	2.865
Diameter (μ)	0.1	10	0.1
Aspect ratio	Large	>100	50 to 100
Surface area ($cm^2/g \cdot 10^3$)	20	—	5 to 14
Habit	Fibrous	Acicular	Fibrous
Hand	Soft	Prickly	Soft
Tensile strength ($lb/in.^2 \cdot 10^5$)	2 to 10	2 to 3	1.5 to 3.3
Young's modulus ($lb/in.^2 \cdot 10^6$)	20 to 25	6 to 16	13 to 21
Temperature range (°C)	400 to 600	970	770
Hardness (MOH)	3 to 5		4

those who will wish to relate the fiber to "Brown Lung," the sickness similar to asbestosis suffered by textile workers. But these people will manufacture some type of scare tactics without regard to facts or needs.

6.8. Tests

When considering tests, one is at once confronted with units. It is my feeling that most readers will probably be more comfortable with the more familiar units, such as pounds per square inch, than with giga newtons per square meter or giga pascals. I plan to include a short table at the end of this chapter, and those who have a compulsion to remain up to the minute in all their calculations may feel free to use any units they choose.

6.9. Properties of Phosphate Fibers

Table 6.1 is a compilation of properties of phosphate fibers as compared to chrysotile.

There are many properties of phosphates that have respectable values when compared to other competitive fibers. Chrysotile was chosen for comparison, because it was this area that was most likely to be a one-on-one replacement in the early days of commercialization.

6.10. Conclusion

Conventional condensed phosphate chemistry was used to prepare phosphate fibers, with some additions designed to cause polyphosphate molecules to grow

Table 6.2. Units of Conversion

To convert from	To	Multiply by
Pounds\in.^{2}	Newtons\m^2	$6.8947 \cdot 10^8$
Pounds\in.^{2}	Pascals	$6.8947 \cdot 10^8$
Pounds\in.^{2}	Grams\cm^2	70.307
Newtons\m^2	Pascals	1.00000

extremely long. This was accomplished by growing crystal in melts in which the metal oxide chain terminators would not block growth of molecules on seed crystal templets.

An error was corrected in the $Ca(PO_3)_2$–$NaPO_3$ phase diagram resulting in the discovery of the $[NaCa(PO_3)_3]_n$ phosphate fibers. These were the fibers that were commercialized and were more easily handled than the $[Ca(PO_3)_2]_n$ fibers, that were of more interest in areas where bone prostheses were being prepared and tested.

It had been shown that it was possible to grow fibers from food grade materials, currently eaten in many foods, and to prepare cloths that cannot burn. This project was never completed, but the work required to make a cloth with pleasing properties should not be a gigantic project.

6.11. References

1. E. J. Griffith and R. L. Buxton, *Inorg. Chem.* **4**, 594 (1965).
2. E. Thilo and U. Schulke, *Angew. Chem.* **75**, 1175 (1963).
3. J. R. Van Wazer and S. Ohashi, *J. Am. Chem. Soc.* **80**, 1010 (1958).
4. W. Wieker, A. R. Grimmer, and E. Thilo, *Z. Anorg. Allgem. Chem.* **330**, 78 (1964).
5. D. R. Gard, *Phosphorus, Sulfur, Silicon* **76**, 231 (1993).
6. O. T. Quimby, *J. Phys. Chem.* **58**, 603 (1954).
7. E. J. Griffith, *J. Pure Appl. Chem.* **44**, 173 (1975).
8. R. K. Osterheld and R. P. Langguth, *J. Phys. Chem.* **59**, 76 (1955).
9. E. J. Griffith, *J. Inorg. Nucl. Chem.* **26**, 1381 (1964).
10. G. W. Morey, *J. Am. Chem. Soc.* **74**, 5783 (1952).
11. J. C. Gremier, C. Martin, and A. Durif, *Bull. Soc. Fr. Mineral. Crystallogr.* **93**, 52 (1970).
12. Y. B. Peng and J. B. Koen, *Bioabsorbable Glass Fiber/Polymer Composites*, MO-SCI Corporation, Rolla, Missouri (June 30, 1993).
13. R. Pfansteil and R. K. Iler, *J. Am. Chem. Soc.* **74**, 6059 (1952).
14. E. J. Griffith, *American Symposium of Phosphorus Chemistry*, ACS Symposium Series 486, Eds. E. N. Walsh, E. J. Griffith, R. W. Parry, and L. D. Quinn, American Chemical Society, Washington, DC New York (August, 1991).
15. E. J. Griffith and I. J. Kodner, U.S. Patent No. 3,312,523 (April 4, 1967).
16. E. Thilo, *Chem. Technik* **4**, 113 (1952); K. Dornberger-Schiff, F. Liebau, and E. Thilo, *Acta Cryst.* **8**, 752 (1955).
17. M. M. Crutchfield, Unpublished observations.
18. E. J. Griffith, U.S. Patent No. 4,346,028 (August 24, 1982).

7

Scale-Up from Laboratory to Plant

Very often scale-up from laboratory to pilot plant or full-size manufacturing plant can present many challenges that are not easily resolved. It is in this domain that chemical engineers make their greatest contribution to new progress. Chemists, more than engineers, are inclined to be artistic rather than practical. They often build art-forms of science for the archives rather than useful products for the everyday world. (I take license in this statement because I am a chemist and not an engineer.)

Reactions that are simple in a laboratory can be too expensive to operate on a manufacturing plant scale. When expense and profitability are of no concern, anything that can be done in a laboratory can be done in a plant. When constraints of a practical process are overpowering, a project may have to be abandoned. If a process can be implemented in a manufacturing plant, the manufacturing plant will usually prepare products superior to those made in a laboratory. Without the engineer's skill and knowledge most new processes would never function properly or become profitable.

Chemists usually have much latitude in exploring new variables in a laboratory. A key point to the successful scale-up of a commercial process is to keep it simple. Not only the chemistry, but also the equipment should be as simple as is practical while being safe. Contrary to popular belief among business management, laboratory research is relatively inexpensive. Managers like to see wheels rolling, people scurrying in a burst of activity, while a product is leaving the end of a tube at a ten to one-hundred million pound per year rate. Anyone can see that this is progress.

The toys used in the laboratory are small and make very little commotion. Management is never truly certain what is happening. Too often, management does not understand the technology they are selling and they are intimidated by the science they do not understand. This is despite the fact that they often speak the language without an in-depth understanding of the concepts.

7.1. Research versus Development

It is the development of a new product that is expensive, not the research required to define a new product. It is expensive to experiment with pilot plant or manufacturing plant sized equipment. In a laboratory, if some part of a process does not work properly it requires very little effort to make changes to try something new or different. Development in a plant, even a small pilot plant, requires a crew to turn the plant around. If new equipment is needed it is usually expensive, and if an out of specification product is made during an experiment, it must either be reworked or disposed of in some way. These disposals are always expensive when working under the regulations of today. It costs much more to dispose of a chemical than it does to purchase the chemical. Much research that should be done is pushed aside because of the extensive paper work required to dispose of unused chemicals.

Unfortunately, development is always lumped under the heading Research. It is characteristic of business management to wish to push a baby project from the laboratory before it has completed a reasonable gestation period. This is both good and bad. It is good that the project is not being ignored. It is bad because the projects are seldom ready to develop and research is conducted in larger-scale equipment rather than laboratory equipment. A classic example of a few years ago was a plant to make phosphorus. It was finally abandoned at great expense. Some laboratory work should have shown that it was a very poor design.

7.1.1. Engineers Do Their Thing

Engineers are already confined to a number of parameters that are very often difficult to change if a product defined by chemists is to be manufactured. Unlike chemists, engineers are attempting to do exactly the same operation, over and over again, exactly as was done when a superior product was obtained. This is one reason products produced in a manufacturing plant are usually superior to those produced in a laboratory. There are many other reasons why plant products can be better than those produced in laboratories. When huge quantities of materials are produced, it often limits the amounts of impurities that can get into a product. Surface-to-volume ratios of a product made in laboratories are usually much greater than those produced in manufacturing plants.

Assume that a product, calcium polyphosphate for example, has been prepared in the laboratory and possesses properties that are acceptable for a marketplace product. This judgement is based on tests performed in the laboratory and potential customer sampling. In the laboratory the product has been made from Reagent Grade monocalcium orthophosphate that was carefully mixed with Reagent Grade 85% orthophosphoric acid. In a manufacturing plant the raw materials will not be Reagent Grade and it is very likely that the engineers must move one step back in the process and use lime and phosphoric acid as starting materials rather than monocalcium orthophosphate and 85% phosphoric acid. Nevertheless, when start-up is completed and the plant has been lined-out, a superior product will be made almost every time.

7.1.2. Analysis of Raw Materials

The monocalcium phosphate used in the laboratory was analyzed for Ca, P_2O_5, and weight loss on ignition by the zinc oxide method.[1] Reagent Grade monocalcium orthophosphate is truly monocalcium orthophosphate only by mistake. Monocalcium orthophosphate is unstable and will disproportionate into a mixture of orthophosphoric acid and dicalcium orthophosphate given any opportunity. The reagent will contain either an excess of orthophosphoric acid or an excess of dicalcium orthophosphate to stabilize it. Van Wazer published a phase diagram based on several literature works that points up the reasons for the instability of $Ca(H_2PO_4)_2$ and its hydrates.[2] In either event it is imperative that the reagent be analyzed before it is used. The orthophosphoric acid was analyzed for P_2O_5 and weight loss on ignition by the zinc oxide method mentioned above.

After being analyzed, monocalcium orthophosphate was weighed on a laboratory balance that was capable of weighing to 0.01 g. The weight of the monocalcium orthophosphate was recorded in a notebook and then the correct weight of analyzed orthophosphoric acid was added to the monocalcium phosphate. This mixture was stirred well and then transferred to a 500-ml platinum crucible. The crucible is placed in a furnace at some temperature high enough to drive off all free water. If this slurry is not first heated at lower temperatures, the charge will splash product over the walls and floor of a furnace. If phosphate melt is heated on a laboratory furnace wall, most furnaces soon become useless. The liquid melt can penetrate the walls, allowing the phosphates to react with the ceramic's walls. This usually causes the furnace walls to crack when the furnace is cooled and reheated.

7.2. Crystallization Variables

In Chapter 5 the study of melts and crystallizations were considered in detail. In the Phosphate Fibers Project the controllable variables of crystallization were limited. After the project was underway two candidates were chosen. All other fibers were given very low priorities. Once the growing medium for $[Ca(PO_3)_2]_n$ crystals was specified, only doping salts were studied as composition changes and these were minor constituents.

The maximum temperatures to which melts should be heated and the length of time the melts should be held at maximum temperatures is important. If the temperature is too high above the melting temperature of the crystals of $[Ca(PO_3)_2]_n$, volatiles other than water, such as P_2O_5, will begin to escape from the melt, changing the composition of the melt. At higher temperatures any residual seed crystals left in the melts will be destroyed. This will help to keep extraneous seeds from influencing the melts and is a positive influence. A balance between heating the melt hot enough to give the desired properties and cool enough not to volatilize product is desired.

Rates of cooling are very important. If melts are cooled too rapidly, either glasses form or very small ill-formed crystals are usually grown. When glasses are first formed and then crystallized as glasses, the crystals derived from glass usually have different habits from those grown directly from melts, where the liquid is fluid.

When a glass-forming type of melt is cooled, these systems are inclined to supercool, unless they are cooled slowly and are seeded with active seed crystals. This means that Newton's law of cooling applies to a supercooled melt system. Newton's law states: The degree of superheating of a system when crystallizing is directly proportional to the degree of supercooling the melt encountered before it began to crystallize. Newton's law is very important in all types of thermal analyses where measurements of glass formers are made. It is also important during crystal growth in both laboratory and industrial processes.

Ideally, a melt should be cooled to a temperature near the melting point of the system. In this case the melting temperature is the temperature at which the last trace of crystal in a melt disappears as the melt is slowly heated. Now when the melt is seeded and slowly cooled, crystals will grow slowly and they should have a good form. Crystallization should follow the temperature–composition curve of the phase diagram of the system. If the temperature is lowered too rapidly, crystallization will lag behind and the melt will supercool. If the temperature is held too close to the composition–temperature curve, crystallization will be too slow to be practical. Crystals that grow on the surface of a melt cool first and grow more rapidly. Surface crystals will usually be unlike crystals grown in the bulk of a melt.

Mechanical agitation is yet another variable that is important during the crystallization of a melt. If a melt that is difficult to crystallize is stirred mechanically at a temperature slightly above the melting point of the crystals, crystals will form very rapidly as the system is cooled, but the crystals are usually extremely small. Large crystals of good form and habit are more likely to be grown from quiescent melt systems.

Yet another variable that cannot be ignored is the capacity and shape of a crystallizer. Heat transfer from a system will depend upon the quantity of melt being crystallized, the surface-to-volume ratio, the temperature gradient between the melt and its environment, the heat of crystallization of the crystals, and the viscosity of the melt, to mention but a few that could be considered. Impurities leached from a container can also have great influence on the behavior of a system. Some substances can act as threshold agents in melts, just as polyphosphates act as threshold agents in aqueous solutions, preventing the crystallization of scale in boilers, sewers, etc.

This section has considered the growth of $[Ca(PO_3)_2]_n$ as grown from a melt. If $[NaCa(PO_3)_3]_n$ is grown from a melt, most of the same considerations must be included. It will be seen that melt chemistry was finally avoided during the growth of $[NaCa(PO_3)_3]_n$ crystals, yielding a practical method for preparing the fibers.

7.2.1. Seeding

When it was evident that most of the free water had been expelled from the phosphates in the melt preparation of $[Ca(PO_3)_2]_n$ above, the furnace temperature was increased slowly to perhaps 1000 °C. At this temperature the melt is an almost white hot liquid with a viscosity similar to syrup, as judged by the way it flows in a platinum dish when the dish is tilted. After perhaps 30 minutes at 1000 °C it was judged that the product was sufficiently free of water of composition to lower the temperature of the furnace. The furnace temperature was lowered to about 900 °C, where the melt was seeded with crystals from a previous preparation if these seed crystals were available or existed anywhere. Crystallization usually commenced immediately, unless the temperature was still too high or the melt had not yet reached the desired composition.

7.2.2. Other Variables of Scale-Up

After the melt was seeded, any number of variables must be considered in ramping down the temperature of the furnace to crystallize the melt. As mentioned, slower crystallization rates yield larger and more nearly perfect crystals, but if time is excessive no production rate would ever be achieved. A compromise between quality and production must be resolved. This will require estimates of the size of the market to be served, an expected sales price, and capital available for a plant, because slower production rates will require larger, more expensive plants.

Two other highly significant variables must be considered. Is the product to be crystallized as gigantic single masses or as many smaller masses? In the first case surface-to-volume ratios become critical. Large masses contain much more total heat which must be dissipated through a smaller surface that is nearly always crystallized first. Heat in the melt must transfer through a surface that is unlike the remainder of the charge. Here a melt will usually cool and crystallize at its own rate without being influenced greatly by outside environments, provided the melt is protected from outside environments. Heats of crystallization are heating the mass while heat is being dissipated through its crystallized surfaces.

7.2.3. Surface-to-Volume Ratios

If crystallization is to take place in many smaller containers, then container shape and dimensions begin to become important. A shallow dish will cause a melt to crystalize more rapidly than a deeper container. This introduces yet another, not so obvious variable. Crystal quality and length will depend upon container dimensions. Fibrous-type crystals tend to grow perpendicularly to the top and bottom surfaces of deeper melts, while growing parallel to surfaces of shallow melts.

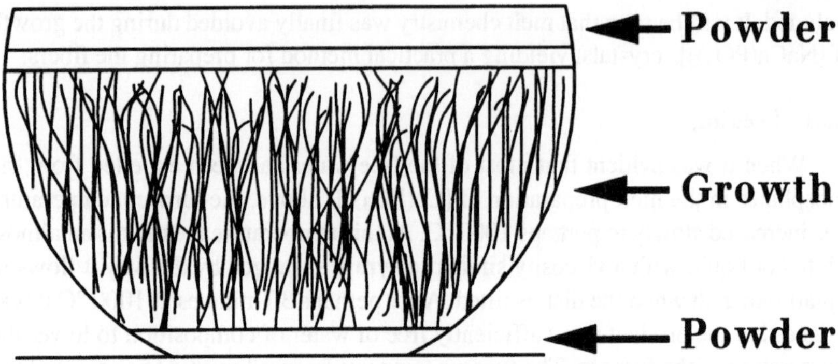

Figure 7.1. Side view of a fibrous crystal grown in a crucible.

Crystals that grow parallel to a surface are usually longer, but of poorer quality, because they crystallize more rapidly and because they grow in a two-dimensional space. Gravitational influences at a surface influence crystals that grow parallel to a surface, because the crystals are asymmetrically supported by the melt. This also depends upon the size and depth of the container. In laboratory vessels, crystals that grow perpendicular to top and bottom surfaces are usually contained between layers of materials that are dissimilar to the mass of crystals being grown. Impurities are usually, but not always, pushed to the top and bottom of the container. Figure 7.1 shows the influence of containers on the growth of fibrous crystals. The same behavior is observed in large melts.

7.3. Two Approaches to Crystallizing Fibers

Two entirely different approaches to the growth of fibers were required for $[Ca(PO_3)_2]_n$ and $[NaCa(PO_3)_3]_n$ fibers. Calcium polyphosphate fibers were grown as dendritic spherulites, while sodium calcium polyphosphates were grown as single crystals that were not fibrous in any respect until milled. Both crystals could be grown in a belt furnace discussed below, but only the $[NaCa(PO_3)_3]_n$ crystals could be grown in a calciner in which feed crystals were sprayed with feed liquor. With additional work it is possible that even the $[Ca(PO_3)_2]_n$ could have been grown in this approach, but the decision had been made to commercialize $[NaCa(PO_3)_3]_n$ crystals rather than calcium polyphosphate crystals. Both approaches will be discussed in detail.

7.3.1. Scale-Up of Calcium Polyphosphate Fibers

Scale-up from laboratory to a modest pilot plant scale was no minor task for preparations of calcium polyphosphate samples of 50 to 100 lb per batch. Containment problems suddenly take on a new complexion. To lose control of 100 lb of molten phosphate at temperatures approaching 1000 °C could cause major prob-

Figure 7.2. The furnace used to prepare large samples of $[Ca(PO_3)_2]_n$ fibers.

lems for both equipment and personnel. Unless protected well, the breakage of a ceramic crucible could destroy an expensive furnace and cause very dangerous conditions if it suddenly spilled onto a worker's feet. See Figure 7.2.

To minimize the damage that could occur to a furnace and danger to personnel in the event of a spill, the furnace floor was covered with a course sand to absorb as much of a spill as possible. A bed of sand was also maintained at the base of the furnace to trap any spillage, keeping it from flowing. Fortunately, this never became a problem. Several crucibles cracked during their use, but none catastrophically. Since crucibles were intentionally sacrificed upon the completion of crystallization, cracked crucibles presented no significant problems.

7.3.2. New Meaning of Product Specifications

When attempting to research and develop a product that is to be marketed for its solid physical properties rather than its chemical properties, consistent quality becomes of paramount concern. This is so because the products used for their physical properties can cause catastrophic accidents if they fail unexpectedly. Small changes in process variables can cause large and even devastating changes in product quality and functionality.

Consider the brake pads on your automobile as an example. You cannot accept a sudden failure of your brakes during a panic stop. Manufacturing customers cannot tolerate unexpected changes in quality. Their product may fail, if their raw materials cannot be depended upon day after day after day. Customers cannot tolerate even an *improvement* in their raw materials without extended prior knowledge of a change. No improvement can be made without change. Nothing can remain the same and improve simultaneously, despite the ill-fated dreams of those desiring to return to nature while maintaining all *current* advancements. An improvement is an out-of-specification product unless customers have had the opportunity to adjust their processes for an *improvement*.

7.3.3. Potential New Customers Come to Life

In order to make samples of phosphate fibers for supplying potential customers, laboratory chemists under the direction of Dr. Richard Hansen had initiated scale-ups to prepare 100-lb samples routinely. These operations were satisfactory for the purpose they served, but containment was a problem. Crucibles used to prepare melts were sacrificial because melt and crystals would grow to the walls of their containers. It was obvious that a manufacturing plant could not destroy a large part of its equipment for each batch of product produced.

7.3.4. How to Crystallize Tons of Melt

Even before engineers were requested for help, all manner of schemes were proposed for potential ways to overcome corrosion and methods to remove a charge from processing equipment, when a melt had crystallized in a vessel, making itself a part of the container walls. One suggestion was that a large hole be dug into an appropriate place in earth. Molten phosphate could be poured into this hole to cool. When melts had crystallized, phosphate fibers could be mined from this hole much as chrysotile is mined. This idea was to emulate nature on a much smaller scale. Other ideas included the use of slag pots or ladles that could be opened as a clam shell. Every idea was given consideration. Most were impractical for one reason or another.

Before much engineering work was needed, it was required that a product be defined. There were several candidates. Two leading candidates were calcium polyphosphate and sodium calcium polyphosphates. Calcium polyphosphate was more difficult to make. There was probably no way that a product of this type could be made without heating raw materials to a clear melt at temperatures near 900 °C. Yields were not as high as with sodium calcium polyphosphate. An overriding property of calcium polyphosphate, making it a poor second choice, was its irritating prickly nature.

One young chemist was heard to say, "If I wanted to torture a person, I would strip him and lock him in a room of calcium polyphosphate fibers." His point was well taken and this problem has never been solved, but it probably could be solved in a relatively short period of time. It should be emphasized that only $[Ca(PO_3)_2]_n$ fibers, and not $[NaCa(PO_3)_3]_n$, were irritating. Not all batches of $[Ca(PO_3)_2]_n$ fibers were irritating. This alone is evidence that this problem can be eliminated. It is merely a matter of knowing which variables must be controlled. Some work

Figure 7.3. Photograph of the $[NaMg(PO_3)_3]_n$ salt. This salt crystallizes as soft lustrous fibers.

indicated that magnesium phosphate doping could eliminate the problem. Later work with $[NaMg(PO_3)_3]_n$ produced very soft fibers that were similar to $[Ca(PO_3)_2]_n$ fibers in form, but had no prickly nature. See Figure 7.3.

Not only the containment problem was difficult with calcium polyphosphate melts, but leaching and fiber opening were also involved. Separation problems of leachate solubles were certain to be difficult engineering challenges. It would be necessary to recapture dissolved phosphate in leach water before it could be recycled. The cost of phosphate in leach water was too expensive to consider it as waste and recycling was to be expensive also.

As problems were identified, solutions would arise almost as if by magic. Nevertheless, it was obvious that calcium polyphosphate fiber could never be commercialized as a commodity product using this process. It still has some very interesting properties for specialties where price is not a major issue. Some medical uses could fall in this area. Abe's work in Japan has concentrated on this particular compound and some outstanding results have been obtained.

7.4. Melt Containment

At the time scale-up was first being considered seriously, both calcium polyphosphate and sodium calcium polyphosphate were still being crystallized from high-temperature melts. It was certainly not practical to consider making these salts in platinum crucibles as was done in the laboratory. Everything that had been tested was soon destroyed by molten red hot phosphate. Some work had been done with phosphates in graphite crucibles in high-temperature furnaces in which air from the laboratory was allowed to enter. These experiments were a disaster. Crucibles were unusable after being used only once. It was noted that crucibles were destroyed more by air than by phosphates, but there was some phosphate attack under these conditions. This experiment was put aside with other failures.

7.4.1. Graphite Dishes

As failure after failure was pushed aside, someone decided to attempt the graphite experiment again, but in a furnace where contained gas was nitrogen, while air was carefully excluded. The Phosphate Fibers Project had become an outstanding team effort. In a team effort, one idea builds on another until it is often difficult to determine who first discovered a concept. Development works well with teams. Research seems to be much more of a "loner's game." It is usually some one identifiable individual who has a first idea and does the first experiments to prove an idea. Whomever it was, someone first considered using an atmosphere of inert gas about a graphite crucible containing a molten phosphate, tried it, and it worked splendidly. Not only was the crucible hardly attacked, but crystallized products fell from the crucible with no signs of sticking.

It seems to be an unwritten law of nature, "Nothing is ever absolutely perfect." Some something will sneak into the best that humanity is capable of doing and put some blemish upon it. Carpet weavers have known this for generations and build a slight error into every luxury rug, in hopes of eliminating larger unwanted errors. Graphite containment vessels were no exception. When elemental phosphorus is manufactured, it is carbon that is used to reduce P_2O_5 to elemental phosphorus and carbon monoxide. It should have been expected that there might be some small amount of this reaction in a carbon crucible.

In manufacturing elemental phosphorus some form of SiO_2, usually quartzite, is added to calcium phosphate, apatite, to drive P_2O_5 from the calcium phosphate. Calcium silicate is more thermodynamically stable at higher temperatures than is calcium phosphate. It is the author's belief that liberated P_2O_5 is then reduced by coke in a furnace burden to yield gaseous elemental phosphorus. Elemental phosphorus is distilled from a furnace and condensed much as distillate in any other distillation process. This accounts for the very high purity of products made from elemental phosphorus. Phosphorus has been distilled and is itself of very high purity.

Crystals of phosphate fiber that were crystallized in graphite crucibles contained a slight pink discoloration in an area of fiber cake that contacted crucible walls. It was not extensive because there was little to drive a reduction, since neither CaO or Na_2O had counteranions such as SiO_2 to drive P_2O_5 from a hot melt. For whatever the reason, probably a slight unbalance of charge, some red phosphorus seemed to be forming in these cakes of fiber.

7.4.2. A Belt Furnace is Successful

Once containment problems had been solved to a degree that it was not a major complication, attention was directed toward the kind of furnace that might best be suited to take advantage of this new-gained knowledge. The first furnaces were expected to produce about ten million pounds of product per year when phosphate fibers were to be introduced into the marketplace. With a completely new product it is always wiser to start rather modestly, because one can never be completely certain how events will shape a product's future, nor how customers may react after once having thoroughly tested their markets.

Engineers decided to purchase a long metal belt furnace that was capable of reaching the required temperatures and could be continuously purged with nitrogen. This furnace was approximately fifteen feet long and two and one half feet wide. It was slotted at an entry end to allow charged dishes to be fed onto a belt automatically. It was soon learned that seeding was as important to production as it was to research. An automatic seeder was built that seeded each dish of melt at the precise moment it was required.

Everything about this furnace (Figure 7.4) performed as well as or better than expected, but now we are back to the containment problem. What style and shape

Figure 7.4. The belt furnace used to prepare larger samples of $[NaCa(PO_3)_3]_n$ fibers. Reprinted by courtesy of the Harper Electric Furnace Corp.

Figure 7.5. Graphite dish used to grow crystals in the belt furnace.

should a graphite crucible be to produce about 1,000,000 pounds of phosphate fiber per year per furnace? This solution was much simpler than I had expected, again bringing to light the advantages of working with exceptional engineers. The product was to be contained in graphite dishes (Figure 7.5).

These dishes were 8" × 8" × 2.5" squares with 1/2" walls and bottoms. Top surfaces of melts were left open to the furnace atmosphere of nitrogen. If the dishes were not contained in an inert atmosphere they would burn at melt temperatures. The furnace belt could be moved at a rate that was preprogrammed, but samples had to move at a fixed rate throughout their trip in the furnace. Temperature profiles could be adjusted at will. In early parts of a sojourn the furnace was programmed for temperatures near 1000 °C. When samples had moved far enough into the furnace to be certain that melts were formed, cooling was begun. Most water of composition had been removed before temperature profiles were adjusted such that samples could cool at an optimum crystallization rate. Crystallization was completed just before samples emerged from the furnace. Samples were dumped into a bin and then transferred to a crusher. This was followed by a mill if sodium calcium polyphosphate fibers were prepared.

7.5. Can We Grow Fibers in a Calciner?

The system discussed above performed well and made a good product, but it was almost certain to cause problems as a manufacturing unit. There had to be a better way. This is when engineers decided to attempt a larger calciner, based on information learned in the laboratory as discussed in Chapter 6. First results were not very encouraging because some severe corrosion problems developed.

When the research team learned of these corrosion problems, they were recognized as the same problems encountered in laboratory calciners. To avoid corrosion, calciners had to be operated at higher temperatures. It seems as though this is the wrong thing to do, but when they tried higher temperatures, almost in desperation, corrosion subsided to a normal expected rate for most calciner operations. The reasons for this were discussed in Chapter 6.

From Laboratory to Pilot Plant

This work started in a laboratory calciner as described in Chapter 6. It was demonstrated that $[NaCa(PO_3)_3]_n$ fibers could be prepared by spraying a solution of orthophosphates onto a bed of hot crystals. It is an entirely different problem to design a plant capable of preparing millions of pounds per year. Even room-temperature vessels that contain noncorrosive solids or liquids present new problems that lie almost entirely in the realm of engineering.

To bring the problems into perspective for those who have never faced this design, let us proceed from the beginning through a part of what is required. It was desired to build a plant that could produce at least 1,000,000 pounds of $[NaCa(PO_3)_3]_n$ per year. To start, it is known that four major raw materials will be required. These are sodium hydroxide, quick lime, phosphoric acid, and water. The quality of sodium hydroxide and lime can vary depending greatly upon the supplier. Cooking lime and slaking lime to prepare calcium hydroxide is an involved art and science itself. One does not merely take a bottle labeled CaO from a shelf, pour it into a beaker, and expect to obtain satisfactory results consistently.

7.6. Raw Materials

The list of raw materials that could be used in this process is extensive. Soda ash or Trona could have been substituted for sodium hydroxide. Under some conditions limestone, $CaCO_3$, could be substituted for quicklime, CaO. There are several grades of phosphoric acid that might have been used, but I always insisted upon food grade materials even though they cost somewhat more than industrial grades. It is also possible to use sodium phosphates as monosodium orthophosphates for the sodium hydroxide and phosphoric used in the first plant. And, of course, it was possible to use calcium phosphates such as monocalcium orthophosphates. One requirement was paramount. If uniform melt was to form on the surface of a growing crystal, the feed solution must be uniform. The simplest way to accomplish this was to have a soluble feed and to spray a clear solution onto the bed of hot crystals.

7.6.1. Lime

To grow phosphate fibers that possess strength and consistent properties requires controlled raw materials. In view of the chemistry discussed in previous chapters, it is evident that if extremely long polyphosphate anions are to be grown from a melt, purity of raw materials is critical. Lime is the most unpredictable chemical in the process. Every unplanned flaw in a crystal lattice produces an inferior fiber, often adding to undesirable solubles in a product.

Quicklime is one of the more useful and chemically difficult of any large volume chemical.[3] Difficulties can begin at the mine where calcium carbonate is obtained. Some deposits of limestone, calcium carbonate, are very difficult to convert to lime at economical temperatures. Usually, the more marbleized limestone becomes, the more difficult it becomes to burn it to quicklime.

$$CaCO_3 \overset{\text{Heat}}{\Rightarrow} CaO + CO_2 \tag{7.1}$$

Some preliminary data obtained in our laboratory indicated that limestone containing halides, chlorides in particular, were more easily converted to quicklime than were limestones containing little or no halides. The data suggested that limestone deposited in salt or brackish water may be more easily burned than $CaCO_3$ obtained from fresh waters.

7.6.1.1. Vertical versus Rotary Kilns

Some sources of limestone are more than 99.8% pure $CaCO_3$ and are one of the least expensive items in commerce. Freight can be almost as expensive as the product itself. "Cooking" limestone can provide a very wide range of useful properties. Limestone is usually cooked in two types of kilns, rotary and vertical. Most quicklime is obtained from rotary kilns and is the poorest grade of lime. It is usually made by mixing limestone with coal and then firing the two together in a kiln similar to a Portland Cement kiln. The product contains the ash from the coal and all its impurities as well as the impurities from the limestone itself.

Purer quicklime is obtained from vertical kilns. In vertical kilns limestone is fed into the top of the kiln while fuel is introduced at the bottom. Either fuel oil or natural gas is used to cook the limestone. Fuel oil usually contains enough sulfur to generate SO_2 and SO_3 that are trapped in the quicklime, in much the way these gases are removed from stack gases in powerplants. Natural gas is the highest grade fuel and produces high quality lime. In my opinion it is the only lime that should be used as a raw material in chemical processes where it is to become a part of a consumer product.

There are at least two other critical variables that should be considered when purchasing lime for a chemical plant. Quicklime is usually classed as soft burned, medium burned, and dead or hard burned. The higher the temperature limestone is converted to quicklime, the less reactive it becomes. It can become almost impossible to slake (hydrate) quicklime to calcium hydroxide, if it is cooked at too high temperatures. Cooking limestone is an art and operators become very skill at their trade.

7.6.1.2. Slaking Lime

As limestone is cooked to higher temperatures, the CaO crystal lattice shrinks and the quicklime becomes more and more dense. Soft burned lime is more reactive and can be slaked more easily, but slaking is yet another art form. Other than the source and history of a quicklime, the temperature and concentration of a lime slurry become important variables. In chemical processes it is usually desired that a slaked lime react quickly and completely with other raw materials in the process, such as phosphoric acid. Slaked lime is very sparingly soluble in water and is usually introduced into a process as a slurry. Small particles of lime usually react more rapidly, but not always.

7.6.1.3. "Hexcrystals"

Care must be made in preparing slaked lime, $Ca(OH)_2$, that "hexcrystals" are not formed in the product. "Hexcrystals" are beautiful, almost perfect, hexagons, but they can be troublesome. Often they do not react within the sojourn time of many processes. I have observed hexcrystals that first failed to react with phosphoric acid in a reactor designed to manufacture dicalcium orthophosphate, $CaHPO_4$, and then failed to react in a calciner fired at 400 °C to produce calcium pyrophosphate, $Ca_2P_2O_7$. These hexcrystals emerged in perfect condition and could be found in a product in the condition they left the lime slaker. Since hexcrystals will dissolve very slowly they can cause a product to exhibit shelf-life instability. A product may show an upward pH drift if it contains hexcrystals when it is desired that the pH remain constant.

To deal with hexcrystals in my research, I developed two tests that were very useful and were adopted by the plant analytical group when slaked lime caused problems. It is well known that although lime is very sparingly soluble in water, it can be "dissolved" in a sucrose solution. In fact, lime reacts with sugar to form soluble calcium saccharides. This reaction is rapid for properly slaked colloidal lime, but hexcrystals react with sugar very slowly.

The test was very simple. A carefully weighed sample of about 1.000 g slaked lime was stirred for 30 minutes in a 15% solution of table sugar. The lime sugar solution is filtered from the hexcrystals and any other insoluble trash that may have been contained in the lime. The impurities can range from calcium carbonate to sand or metal scraps. The filtered hexcrystals are air dried and weighed to determine the percent hexcrystals in the lime.

If precise data are required, a similar lime sample may be dissolved in a 20% solution of ammonium nitrate. Both lime and hexcrystals dissolve rapidly in ammonium nitrate solutions. This solution is treated similarly to the sample above, except the insolubles contain neither lime nor hexcrystals. These results can be used to correct the data obtained in the sugar solution test.

7.6.2. Sodium Hydroxide

As the phosphate fiber product required more and more fine tuning, raw material quality became a greater issue. Sodium hydroxide is a manufactured chemical and is usually of high quality, though more expensive than soda ash or Trona. Trona is a mined product and is usually less pure than either sodium hydroxide or soda ash. Its formula is $Na_3H(CO_3)_2{\cdot}2H_2O$, a double salt. Trona can be further purified if it is required; sodium hydroxide will probably continue to be the reagent of choice.

7.6.3. Phosphoric Acid

There are phosphoric acids derived from two different kinds of processes. One phosphoric acid is called furnace acid and is prepared from elemental phosphorus. The second source is referred to as wet phosphoric acid, or green acid because it contains enough impurities to be green before it is purified. Wet acid is purified by some kind of solvent extraction process. In general furnace acid is purer than purified wet phosphoric acid, but both can be made very pure.

Elemental phosphorus is made by heating a mixture of phosphate rock, $Ca_4(PO_4)_2(OH,F)_2$ or $Ca_3(PO_4)_2{\cdot}Ca(OH,F)_2$, to write it as the solid solution of tricalcium orthophosphate and calcium hydroxide or fluoride, or both, with quartz and coke. See Chapter 3. Elemental phosphorus is burned to phosphorus pentoxide and the pentoxide is allowed to react with water to form very pure phosphoric acid.

Wet acid is made by allowing phosphate rock to react with sulfuric acid to form calcium sulfate, gypsum, and crude phosphoric acid. Phosphoric acid is extracted selectively into an organic solvent, from which it is stripped to yield a good quality of phosphoric acid. Furnace acid or purified wet acid can be purchased for similar prices. For many uses they are of approximately equal value. I remain biased toward furnace acid when purity is an issue.

Other than a city water supply, no other raw materials are used in the manufacture of phosphate fibers. It is doubtful that this could be a critical issue, but if unexpected difficulties were encountered a water supply can cause some strange problems. An example of such a problem could occur if some unwanted seed crystals should grow from a mineral in a perfectly acceptable potable water supply. This could cause a change in crystal habit and fiber properties.

When electricity, gas, water, lime, sodium carbonate, phosphoric acid, tanks, pumps, switches, valves, analytical equipment, and controls have been properly placed, it is time to consider the kiln itself and all the auxiliary equipment required to run the kiln. Of instant concern today are emissions and waste products. Dust collectors must be in place and operational before a kiln can be operated. There must also be a means of disposing or recycling off specification product. Poor-quality products are usually produced in the first trial runs of new equipment.

When a product exits in the kiln, there must be some means of collecting it and transporting it to a mill where it is reduced to the proper particle size before it is either bagged or shipped in bulk. Either a warehouse must be available, or a transport truck or railroad car must be waiting to receive the finished product, but only after it has been cleared by the Analytical Laboratory. Analytical has the final say as to whether or not a product can be shipped and is within specifications.

At least two more items must be considered. The process cannot be started before Safety has cleared the plant as safe for personnel and the environment. This not only means that it must be safe, but it cannot be a public nuisance for our neighbors. If the plant creates unacceptable noise levels or lighting conditions that can aggravate neighbors attempting to sleep, the plant *will not* pass inspection and will not be started up.

7.7. A Kiln Patent is Granted

Up to this point almost no attention has been paid to the heart of the plant, the kiln. It has several unique features that are unknowns. It is large and must rotate at a high temperature. It is gas-heated internally, but feed to the kiln is a liquid that is sprayed onto a bed of hot crystals through water-cooled feed tubes that reciprocate over the feed to insure that feed is fed uniformly to the hot charge. Perhaps the best way to describe the kiln is to furnish the description given in the U.S. Patent No. 4,717,487.[4]

"Example 1

(a) Crystallization Kiln

A crystallization kiln was constructed from 60.96 cm (24 in.) inside diameter × 121.92 cm (48 in.) long stainless steel tube, including a 5.08 cm (2 in.) long × 13.97 cm (5.5 in.) high discharge chute at each end thereof, having its longitudinal axis oriented along a substantially horizonal line and having a centrally located 33.02 cm (13 in.) circular opening in each end to accommodate supply lines fitted with (a) a 93.98 cm (37 in.) long × 17.78 cm (7 in.) wide × 12.70 cm (5 in.) deep jacketed water-cooled line burner containing two staggered parallel rows 14 and 13 blast burner tips spaced 6.096 cm (2.4 in.) apart, center to center, and having a maximum firing rate 1.90×10^9 joules/hr (J/hr, 1.8×10^6 BTU/hr) of full ratio gas–air mixture at maximum pressure of 2.49 pascals (P, 10 in. column of water) which was positioned in the upper quadrant of the crystallizing kiln shell with its length parallel to the axis of rotation of the shell and the burner tips directed in the direction of rotation of the kiln shell on a 45°

angle from the horizonal and vertical axes into the diagonally located lower quadrant of the kiln shell; (b) a 71.12 cm (28 in.) long × 2.54 cm (1 in.) inside diameter jacketed water-cooled stainless steel thermocouple assembly having four perpendicularly oriented Hastelloy C-276 tipped thermowell arms (non-jacketed tips) at equally spaced intervals along its length [to provide temperature reading and control of each of four equally divided imaginary zones Z1–Z4 of the crystallization kiln shell (numbering sequentially from one end of the kiln shell to the other), one thermocouple for each numbered zone] which was positioned parallel to and just below the line burner to coincide with the 91.44 cm (36 in.) flame zone with the thermowell arms being curved in the direction of rotation of the crystallization kiln shell to run along the kiln shell wall, but spaced from the kiln shell wall at a distance slightly greater than the distance and the boring/scraper bar described in Part (d) below; (c) 91.44 cm (36 in.) long × 2.54 cm (1 in.) jacketed water-cooled stainless steel feed tube assembly containing seven 0.635 cm (0.25 in.) feed tubers numbered F1–F7 branched off perpendicularly to the lengthwise axis of the assembly, all in the same direction, at equally spaced intervals along the length of the assembly and fitted with seven solenoid valves (located externally to the kiln), each associated with a particular feed tube to control the flow of feed material through the assembly, which was positioned parallel to the line burner in the upper quadrant of the annular opening of the kiln shell opposite that of the line burner to coincide with the 91.44 cm (36 in.) flame zone of the line burner and deposit the feed material directly into the flame zone; (d) a 116.84 cm (46 in.) long × 10.16 cm. (4 in.) × 1.27 cm (0.5 in.) thick Hastelloy C-276 bar having 1.905 cm × 1.905 cm (0.75 in. × 0.75 in.) notches spaced 1.905 cm (0.75 in.) edge to edge along its length on one edge thereof which was positioned parallel to the line burner and supported by adjustable brackets attached to the under-side of the line burner such that the notched side of the boring/scraper bar approached the inner wall of the kiln shell [usually about 1.905 cm (0.75 in.) at the apogee (peaks) of the notches and about 3.81 cm (1.5 in.) at the perigee (valleys) of the notches] from the side of the line burner opposite that with the burner tips; and (e) a motor driven rotation drive."

The kiln performed well, particularly as the prototype designed to discover problems in larger designs to follow. It was my belief that larger crystals, with more nearly the exact composition to yield exactly $[NaCa(PO_3)_3]_n$ with the required stoichiometry, should have had less solubles and stronger fibers, but these were design parameters in the work to come. In Chapter 8 it will be seen in some applications, such as automotive brakes, that the solubles played a useful function, acting to behave as a glue to add green strength to air-dried brake blocks.

7.8. Milling

Sodium calcium polyphosphate fibers are grown as small crystals that have little or no similarity to fibers (see Figure 7.6, unmilled). It is only after crystals are milled that they have fiber properties (see Figure 7.7, milled). Naturally, a fiber cannot be longer than the crystal from which it was taken. Fiber length, aspect ratio, and strength are always an issue when working with fibers. If overmilled, phosphate crystals can be converted to a dust. If undermilled, fibers are not well separated and product fibers are inclined to be gritty. If impacted too strongly, part of the crystalline fibers can be converted to amorphous solids.[5] These can cause several

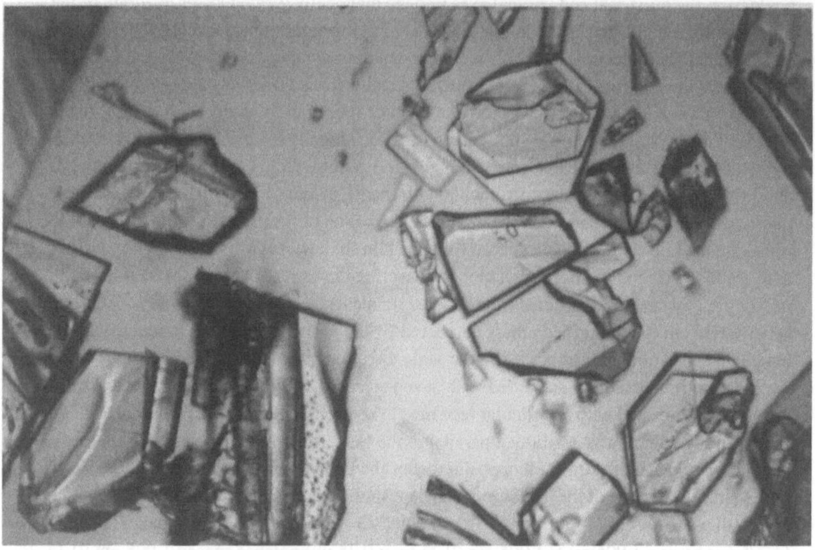

Figure 7.6. Unmilled crystals of $[NaCa(PO_3)_3]_n$.

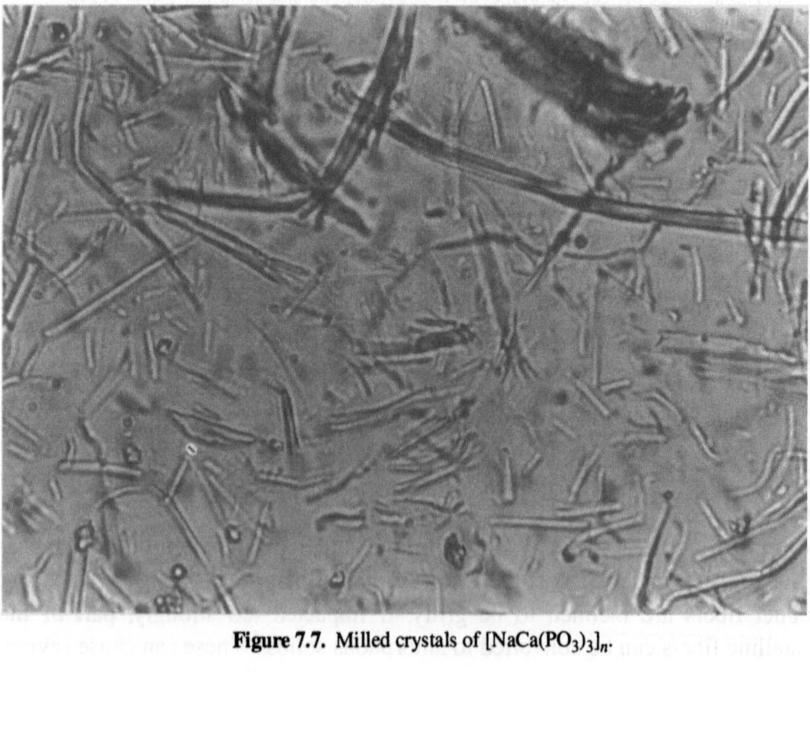

Figure 7.7. Milled crystals of $[NaCa(PO_3)_3]_n$.

problems when using fibers in applications where dissolved solids are unwanted, or where strength or pH control is a requirement. When making inorganic paper from fibers, a latex binder can be sensitive to both dissolved solids and pH.

If pressure is applied to a crystal of sodium calcium fiber, it will cleave into a bundle of tiny fibers aligned with the "*c*" axis of the mother crystal. These fibers are high quality when obtained by crushing, but they are poorly opened. A roller mill should have served this purpose. Some secondary milling operation is required to make an acceptable product. A corn meal type of mill wheel used to create flower and corn meal would probably have worked well, but should be slow since crystals would have required both crushing and grinding. These operations are easily done in the laboratory with a mortar and pestle. First crystals are crushed, and then a pestle is used to grind and open the crushed crystals. This approach was considered to be low technology by most engineers and was never seriously considered.

The approach taken to grind samples in a pilot plant was surely not low technology, but perhaps too sophisticated for the job it was required to do. The mill was an air classification mill. Although this mill functioned reasonably well, it had wear surfaces that were certain to be a source of excessive maintenance when placed in a full production mode. This issue was never faced, but was recognized as a problem that had not been adequately solved. A two-step crushing and milling process should work well.

7.9. Bagging, Shipping, and Handling

Most bagging operations create dust when fine particles are handled. Safe practices require that personnel who are bagging and shipping phosphate fibers should wear protective clothing, and use approved respiration and ventilation equipment. At the product's destination it is also required that safe practices be observed. All dust must be considered undesirable when breathed, and phosphate fiber is no exception in this case.

No special precautions are required to protect the product during transit. It is insoluble, nonhygroscopic, nontoxic, and is chemically stable. Most drums or bulk bins should be satisfactory, since normal care is provided in storage and shipment to insure that the storage area is clean and dry.

7.10. Cost

The price of phosphate fibers and most manufactured substances can never compete with mined products such as "asbestos," nor can it compete in performance in many areas where indestructibleness is a desirable property. Mined products have had to withstand the ravages of time to have survived. They have had centuries to approach a minimum free energy and, until disturbed, they are likely to maintain their current status. This, of course, excludes organic materials that have been protected from oxidizing environments. These substances are usually convertible to inorganic substances such as carbon dioxide, water, and metal oxides.

The safety of phosphate fibers and their unique properties not possessed by minerals would allow them to receive a higher selling price in areas where they can compete without competition from minerals. The food and medical areas are examples of these markets. Very often it is desirable that a product is degradable.

7.11. Conclusion

Scale-up when properly done is a function of engineering and development, and is not a direct part of research. The research work should be completed, or nearly so, before engineering is undertaken. If this is not the case a great job will be done while painting the wrong house. If things do not work out as required it may be necessary to do additional research, but this should be the exception and not a standard operation.

Three approaches were made to identify systems capable of containing phosphate melts. $[Ca(PO_3)_2]_n$ melts were contained in ceramic vessels. This was successful for melts of 100 lb or less, but could not be used for production quantities. Graphite dishes and a hearth furnace were satisfactory for both $[Ca(PO_3)_2]_n$ and $[NaCa(PO_3)_3]_n$ melts, but there are reasons to believe that maintenance may have become expensive.

Raw materials are a critical issue in any process, and were particularly so in the Phosphate Fibers Project since safety was a critical issue. High-purity economical raw materials were readily available at locations where plants had been planned. Lime was likely to be the most tricky to handle, but this is true in most processes where lime is used. It is unlikely that any special problems should have arisen with the phosphate fibers.

When the decision was made to manufacture only $[NaCa(PO_3)_3]_n$ fibers, a calciner approach was the only economical solution to the containment problems. Thousands of pounds of the fibers were produced in a small production unit. When the project was terminated, most of the equipment had performed well and no major problems had arisen that seemed to block the way to a full production plant when required.

7.12. References

1. E. J. Griffith and C. F. Callis, *J. Am. Chem. Soc.* **81**, 833 (1959).
2. J. R. Van Wazer, *Phosphorus and Its Compounds*, p. 515, Interscience, New York (1958).
3. T. C. Miller, *Calcium Oxide and Water*, National Gypsum Company, Buffalo, New York (1961).
4. E. J. Griffith, J. R. Jany, and T. M. Ngo, U.S. Patent 4,717,487 (January 5, 1988).
5. I. Motooka and M. Kobayashi, in *Topics in Phosphorus Chemistry*, Vol. 10 (M. Grayson and E. J. Griffith, eds.), p. 171, Wiley, New York (1980).

8

The Many Uses of Phosphates and Fibers

Almost every week several callers the world over inquire as to sources of phosphate fibers. They ask, "Are you manufacturing this product? Is anyone, anywhere, manufacturing this product?" Regretfully, I must tell them that I know of no commercial source of the fibers. Response to this product has been very rewarding to me, as an inventor. Monsanto's management made a decision not to manufacture the product and to allow their patent estate to lapse, rather than pay annual fees required to keep the estate active. The problems with potential lawsuits are ever present.

8.1. Patent Rights and Applied Chemistry

The concept of applied chemistry is in many ways synonymous with patent rights. Industries or individuals are unlikely to invest vast sums of money into applied research if they cannot expect protection of their inventions once they have been perfected. Most American citizens are unaware that their patent rights were very quietly degraded less than ten years ago. Most citizens still believe that if granted a patent, it is their intellectual property for seventeen years, without them being required to pay fees. This is no longer the case. If a patent is to remain active, fees must be paid and they are not trivial amounts of money.

The bill that created the United States patent system was signed by President George Washington on April 10, 1790. Patents rights are viewed by most believers in the American Dream as fundamental, as basic as free speech and the right to vote without paying a poll tax. It is my belief that patent rights should be maintained and protected for the average citizen even if the average citizen never applies for, or is ever granted, a patent. Less than six million United States patents have been granted since President Washington signed the bill creating the Patent Office. This means that a very small percent of our citizens has ever received patents and most of our citizens have never even seen a patent. Moreover, very few of our scientists have ever received patents.

The right to own patents is a form of freedom; to know that one can apply for a patent, if an invention is legitimate, and that this intellectual property may be held for a period of seventeen years with no fees was a basic freedom. Patents are but one of the fibers in the fabric of freedom and free enterprise. If it is necessary to charge foreign inventors a yearly patent fee in order to maintain a United States patent, then do so, but do not charge American citizens an annual fee to maintain their patents once the patents have been granted.

Under the best of conditions, most patents are of negative financial value to the few of our citizens who may be granted patents. Most average citizens mistakenly believe that when one owns a patent, Easy Street is just around the next corner. This is the rare exception and is very seldom true. Why add insult to the financial injury they suffered to pay an attorney to obtain a patent in the first instance? The United States Patent Office has already extracted its pound of flesh before the citizen is granted a patent, and it costs the Patent Office virtually nothing to maintain a patent once it is granted. Environmental restrictions and new patent laws have all but pushed individual American inventors completely out of the patenting process.

A simple survey of the number of unassigned patents granted each year to United States citizens will reveal that both the number and percent of patents granted to individual United States citizens, not assigned to a corporation or other institution, continues to fall yearly. Gregory in this series, *Topics in Applied Chemistry*, notes while referring to patents granted for ink jet printing devices,

> "Until 1976 the United States was the undisputed world leader in ink-jet technology. However, since then the Japanese have dominated the ink-jet area. This is reflected by the number of patents on ink-jet. For example, in 1984 Japanese companies applied for 145 patents while only two patents were issued to American companies."[1]

In the past the United States fostered, encouraged, and protected participation of common individual inventors, but today regulations and environmental restraints severely limit individual citizens from experimentation irrespective of patent rights.

I requested of Attorney Raymond C. Loyer, Consulting Patent Counsel of Monsanto Company (formally a patent attorney for Xerox), to please present the facts and his expected consequences of the recent changes in our patent laws. Mr. Loyer is much more qualified to make these judgement than I. We are indebted to Attorney Loyer for the following letter:

From: RAYMOND C LOYER
Date: 9/15/94 11:39 PM
TO: EDWARD J GRIFFITH
Subject: Patent Fees

PATENT FEES

American free enterprise and the ability of the individual to invent and develop an invention was drastically affected by the enactment of the "user fee" system for supporting the U.S. Patent & Trademark Office. With enactment of 35 U.S.C. 41 (b), for the first time in U.S. history, owners of patents who are individuals or a small business as defined by the Small Business Administration (13 C.F.R 121.12) must pay a fee in order to maintain their patents in force. This law requires fees to be paid periodically after the issuance of the patent, i.e., \$465 after 3.5 years, \$935 after 7.5 years and \$1410 after 11.5 years. If the owner is not an individual or a small business, the fee is DOUBLE that noted above. The patent owner who fails to pay the fee within six months of the above noted time periods from issue of the patent loses the right to enforce the patent.

The United States was the last major country in the world to require such a fee. It has changed forever the best patent system in the world and has seriously damaged the basis for the technical leadership this country has enjoyed. While the large corporate patent owners can charge off this new "tax" as a cost of doing business, the entrepreneur who is actively trying to interest these large corporations in the invention or is trying to obtain investment capital to develop the invention must be able to pay this fee or lose the patent. It should also be realized that most basic inventions require more than one patent to be properly protected. Only one class of invention is allowed to be contained in any one patent, i.e. compound, process, apparatus, composition.

In most instances, at least some of these "patent taxes" must be paid before the owner realizes any profits from the invention because the early years are usually heavily leveraged in the start-up phase. A good example for cause to be concerned is the invention of xerography. The inventor, Chester Carlson, succeeded in making the invention in 1938 (in his kitchen). This individual obtained about 6 patents while attempting to interest the large corporations in this country already in the office machine business (IBM, SCM, 3M, Kodak, etc.). None saw any potential in the successful development of a practical automatic office copier on the basis of the basic inventions patented by Mr. Carlson. Nevertheless, Mr. Carlson never lost faith in his invention and finally, in 1948, interested Battelle Memorial Institute, Columbus, Ohio in trying to develop his basic inventions. Battelle then obtained capital and further research effort from Haloid Company, Rochester, NY. These organizations labored and invented obtaining numerous patents to protect their investment in research during the next eleven years. On several occasions, Haloid Corporation (now Xerox Corp.) nearly went broke and issued stock repeatedly to stay solvent. Luckily, the people of Rochester NY had more faith in new technology than the large corporations of that time. Had 35 USC 41(b) been enacted in 1942 instead of 1982 it is most probable that office copying would still be confined to Thermafax, Verifax and other non-permanent, inconvenient methods. Secretaries would probably still be using stencils for large numbers of copies from one original. Certainly, Chester Carlson and little Haloid Company would not have been able to maintain in force the numerous patents they obtained long enough for the development that was needed to give the world the first dry writing, automatic office copier.

To add further insult to the spirit of invention Congress also requires the Commissioner of Patents to set fees for services and materials, all for the purpose of supporting the Patent & Trademark Office without funding from the U.S. Treasury. In other words, the Congress has decided that the Patent system is no longer worthy of public support and therefore it must fund itself. On Sept. 1 of each year the Commissioner of Patents must announce the application and processing fees to be collected during the following year thereby revising 37 C.F.R. 1.16 & 1.17. In 1994 the Commissioner has announced a three percent increase in fees over those set for the preceding year. The Small Business Administration rule applies to these fees in the same manner as in post issuance fees discussed above.

In 1993, the basic filing fee for a small entity was $355 plus extra charges depending upon the number and type of claims in the application. Almost anything done in the process of examining a patent application, such as obtaining an extension of time to reply and any petition, now costs a fee. Prior to 1982, when Congress supported the patent system, the basic filing fee for a patent application was $65, extensions of time to reply were free. Who knows how high fees will rise now that members of the bureaucracy are allowed to set them.

Hope the above helps you.

Ray

8.2. Phosphate Fiber Patents

A list of patents that were active or were pending when the Phosphate Fibers Project was dropped is shown in Table 8.1. The two original composition of matter patents that were granted to me, before yearly fees were required to maintain patents rights in the United States, are still active, as relics of the past. A much publicized reason given for abandoning the new phosphate fiber product was a potential for

Table 8.1. The Patent Estate of Phosphate Fiber Patents had been granted or were pending in each of the following nations: Argentina, Australia, Austria, Belgium, Brazil, Canada, France, Great Britain, West Germany, Holland, Italy, Japan, Luxembourg, Mexico, Sweden, Spain, Switzerland, and United States

Title	U.S. Number
Acicular, Crystalline Calcium Metaphosphate	4,360,625
Asbestiform Crystalline Calcium Sodium or Lithium Phosphate, Preparation and Compositions	4,346,028
Friction Material Compositions Containing Phosphate Fibers	Rejected
Process for Preparing Sheet Composites Containing Crystalline Phosphate Fibers	Allowed
Composites of Sulfur and Crystalline Phosphate Fibers	4,484,950
Sheet Composites Containing Crystalline Phosphate Fibers	4,609,433
Process for the Preparation of Asbestiform Crystalline Calcium Sodium Metaphosphate Fiber	4,717,487
Process for the Preparation of Asbestiform Crystalline Calcium Sodium Metaphosphate Fiber	4,729,838
Process for the Preparation of Asbestiform Crystalline Calcium Sodium Metaphosphate Fiber	4,690,809
Process for the Preparation of Substantially Crystalline Calcium Sodium Metaphosphate Fiber	4,725,368
Process for the Preparation of Substantially Crystalline Calcium Sodium Metaphosphate Fiber	4,725,369

United States Patents had been granted on all of the above applications. Some foreign countries were still in contention, but there was little doubt that the patents would be granted in a short time.

unrealistic lawsuits from "asbestos attorneys" who spent their entire careers in "asbestos" lawsuits.

8.3. Patents Abandoned

At the time phosphate fibers were abandoned, risk liability could surely be viewed as a rational consideration. It is now more than ten years since phosphate fibers were first announced. Recently *ChemTech* (September 1993) featured an article in which phosphate fibers were the first item cited as a substance of which society had been deprived, as a result of potential litigations. This was but one of dozens of articles and television programs and newspaper articles devoted to phosphate fibers in recent years. Multiple needs for these fibers still exist and the invention refuses to die. Phosphate fibers cannot be *un*-invented. Judging from current literature, the Japanese are likely to be the first to market phosphate fibers as a new item of commerce and there is every reason to believe it can be a very profitable product for them.

8.4. Uses of Phosphates

As a backdrop for uses of phosphate fibers, perhaps it will be helpful to consider a few of the more important uses of phosphates that directly influence all our lives. Fertilizers have been mentioned numerous times. Little is to be gained by additional attention to fertilizers except to reiterate that, if we are to eat, we shall manufacture phosphate fertilizers in ever increasing quantities, if the population of Earth continues to expand. There is no choice!

Phosphate Detergents

A food-related use of condensed phosphate detergents is to clean dairy farm barn floors and milking equipment to protect the public from infections that could be carried in milk. Nothing else has been found that is as effective. Phosphate detergents are also used to clean and wash hospital linens and operating rooms because no satisfactory replacements have been found. In addition to food-related uses, phosphate detergents are still used in automatic dishwashers for both homes and institutions because no other substances are known that give the public the protection of a phosphate. There is a message here for those who choose to see it.

No one has ever before questioned the need for phosphate detergents in industrial and institutional uses such as restaurants and hospitals, and particularly in automatic dishwashing machines for home use. Hypochlorites are almost universally used to clean tables and counters in restaurants to help us from becoming infected by bacteria left in the foods spilled by other customers. This week a detergent manufacturer is test-marketing a nonphosphate nonchlorine recycled box in a western city. The printing on the box highlights what the box *does not contain.*

It will be interesting to learn how this product performs. If problems do not arise from bacteria left on dishes, it probably will be the first of a new class of green detergents for home use.

The fate of the phosphate detergent must be belabored to place phosphate fibers in a proper perspective. It was the single largest use of phosphates other than fertilizers, and was but a very small fraction of the phosphates used in fertilizers. It is my belief that someday it will become evident that the phosphate detergent did very little harm, if any, and that it is very easy to control phosphate detergents at sewage treatment plants. Even if phosphate detergents were guilty of all the accusations leveled at them, they represent no more than half the phosphates in sewage. Human and animal waste are still contributing much more phosphate to our aquatic environment than critics claim have caused harm to our lakes and streams. The phosphate detergent probably will be reinstated along with modern sewage treatment plants sometime during the first part of the twenty-first century. Better sewage treatment is still needed with or without phosphate detergents.

Tripolyphosphate is Attacked

I was involved in the Detergent Phosphate Flap from its onset. In the summer of 1954, I was probably one of the first industrial phosphate chemists, if not the first, to be informed and confronted by a government laboratory worker, of how phosphates were destroying our water resources. The body of water was the Ohio River and phosphate detergents were less than ten years old. There had not been enough detergent used to harm anything. The men leading the charge against these detergents requested that I visit their laboratory, a short distance across town from the world's largest phosphate detergent producer, "To help us with a problem."

When I arrived at their laboratory and discussed their problem with them, their problem had nothing to do with eutrophication or growth of algae or killing fish. They seemed to be looking for anything that would cause a phosphate detergent to appear harmful. The first of the "biodegradable" detergents was under study and was receiving much scientific and public attention. It seemed to me that these men were seeking any flaw in phosphate detergents they could find. Perhaps they were unhappy with the large detergent company across town or were seeking ways to fund their projects. It was obvious that they were not searching for a project that would point up the merits of phosphates in our environment.

It is almost, if not impossible to get funding for studies of improvements to our environment that come from the use of a product put into the environment. Can you think of one? There are a few, and phosphates are again being added to lakes to increase fish production.[2] It has finally been recognized that if fish are to eat they too must have phosphates.

The only thing the men in the government laboratory in Cincinnati, Ohio could suggest that phosphate detergents might be doing to cause problems was to deflocculate suspended matter in natural waters. They proposed that water companies might have to use slightly more alum to flocculate particulates in potable water, when condensed phosphates were used in detergents. I believed then, and I believe now, that phosphates in our water resources had nothing to do with their initial attack on a large company a few blocks away from their laboratory. They received attention for their attack. Once underway, no amount of data or reasoning influenced those mission-oriented crusaders determined to seek and destroy. Nothing short of complete annihilation was acceptable. They succeeded. You, I, and the remainder of humanity are the losers.

8.5. Food Uses of Phosphates

There are so many properties, other than safety, which make phosphates ideal substances for the improvement of foods that they cannot all be discussed. A few of the more interesting will be highlighted. The acidity of phosphoric acid in soft drinks is known to most persons. The acidity of phosphates as leavening agents is also well known.

Phosphates are used to clean chickens that have been taken to a processing house. Most chicken feathers are infected with salmonella when they arrive at the processing house. Phosphates are used to clean the bacteria from the chicken before it is packaged and frozen for shipment. No other substance performs as well. Fish, on the other hand, retain their just caught flavor and consistency when treated with phosphates when dressed. It has also been demonstrated that long-chain polyphosphates help fish and minnows stay alive in confined quarters. The phosphates help the fish breathe by allowing more water to pass through their gills with less effort by the fish. This has been particularly helpful in keeping fishing minnows alive.[3]

A relatively new phosphate product, SnowFresh®, has done much to replace the sulfites and nitrites that were previously used to keep lettuce and other vegetables fresh after they had been prepared for serving.[4] The product is very effective and has none of the undesired properties of the previously used products.

Back to food processing, phosphate detergents are very effective in removing mold from fresh strawberries and similar fruit and vegetables that must be washed before being shipped to market. The mold count on strawberries is one of the grading characteristics inspected in processing houses. Polyphosphate detergents do an effective job of removing mold from strawberries.

Condensed phosphates are used almost universally in potable water to protect the public health while protecting our plumbing. Phosphates and silicates have been used for many years to control the corrosion of water pipes, thus reducing the quantity of heavy metals and copper dissolved in drinking water.

Dairy and animal products may become rancid by bacterial attack. Calcium and magnesium ions are required for the propagation of the microorganisms and the development of rancidity. If these calcium and magnesium ions are tightly sequestered as phosphate complexes, they are unavailable to the microorganisms.

This allows the product to have a much longer shelf life before becoming unacceptable as a food.

Many salad dressings have phosphates added for a variety of reasons, ranging from emulsification to pH control. Phosphates are among the best known buffering agents and can be employed without adding extraneous chemicals to a formulation. The phosphates are used in instant puddings to cause the systems to gel. Many of the systems use milk sugar and pyrophosphates. Pet foods have phosphates added to give nutrition while helping to clean plaque from an animal's teeth. These same phosphates are used in toothpaste to help fight plaque and tartar on human teeth. This is in addition to the calcium orthophosphate or calcium pyrophosphate that is used as the polishing agent in time-proven tooth-cleaning formulations and paste.

Potential uses of phosphate fibers are almost limitless. Rather than create a long list of products in which these fibers might find economical uses, only those areas where the fiber has a reasonably good chance of commercial success or areas, where they have been tried and have almost no chances of commercial success will be discussed. It should be reemphasized when discussing uses that two different phosphate fibers were tested, acicular $[Ca(PO_3)_2]_n$ fibers and asbestiform $[NaCa(PO_3)_3]_n$ fibers, on which most application work was done. These are but two of a list of ten recognized useful phosphate fibers and perhaps as many as twenty different fibers that are potentially useful, but have not been tested as fibers.

It is probably true that any long-chain polyphosphate can be crystallized as a fiber, if it can be crystallized under most conditions. For example, there is no known way to crystallize long-chain polyphosphoric acid. However, there is no known theoretical reason why this is the case and perhaps polyphosphoric acid will eventually be crystallized. As mentioned earlier, reports of its crystallization could not be repeated. Until now no seed crystals have been found to initiate the crystallization process. Once crystals of polyphosphoric acid are obtained, it is likely that fibers can be grown or even that fibers should be the preferred crystal growth. It is also likely that phosphorus pentoxide could be crystallized as a fibrous substance in much the same way SO_3 is crystallized as a fibrous solid. Crystalline SO_3 has an appearance not unlike crocidolite, albeit much more difficult to handle.

8.6. Cement Pipe: $[Ca(PO_3)_2]_n$

"Asbestos"-cement pipe was the largest single use of "asbestos," any place on Earth. There is hardly a water-sewage system in any technically advanced countries that does not rely on "asbestos"-cement pipes somewhere in their system. If "asbestos" even approaches the dangers it has been reported to be, "The meek shall inherit the earth" sooner rather than later! If I were certain that I could die only from "asbestos" in cement pipe, I should plan to live at least one-hundred-thousand years. Billions of people have drunk water from these pipes for many years with no obvious ill effects. Surely some would have surfaced had there been a problem.

In Europe, "asbestos"-cement pipes were made by mixing fibers with Portland Cement and then allowing the cement to cure at ambient temperatures. In the United States the process was similar, except that rather than curing at low temperatures, pipes were cured in steam pressured autoclaves, at temperatures near 170 °C, for about ten hours. This process makes very high quality pipes. I was astounded that this process could be operated economically, but it was and the product is superior.

Because profit potential from an "asbestos" replacement in the pipe market is huge, much effort was expended attempting to reinforce cement pipes with calcium polyphosphate fibers. It should be mentioned that cement pipe utilized the very poorest, cheapest, and shortest grades of the chrysotile market. It was inexpensive and was not much better than a waste stream of the industry. This made it difficult for a manufactured fiber to compete, irrespective of its safety. It is doubtful that phosphate fibers could be economically attractive even if they performed in cement pipe.

8.6.1. Tests of Cement Pipe

To test phosphate fibers in Portland Cement, fibers were mixed with cement slurry, cast into coupons about 4 mm × 20 mm × 100 mm, and then cured under a variety of test conditions. After curing, samples were subjected to both strength and fracture tests. Although some fibers were capable of withstanding even the harsh treatments to which they were subjected, they did not approach "asbestos" as a reinforcing agent. If they had performed in this function, it is very doubtful that the fibers would have been labile enough to be safe when breathed or ingested. *Safety* and utility were our primary aims, and some utility must be sacrificed for safety.

It was found that phosphate fibers reinforced high alumina cement very well, but this cement is expensive when compared to Portland Cement and it was believed that this market was too small to be considered a viable outlet for fibers. From the onset, the project was developed under a constant threat of withdrawing our funds, if I did not meet some arbitrarily imposed goals before the next meeting with a Business Director. Since the project was being developed long before research was completed, I continued intentionally to ignore deadlines, to the consternation of business-type directors. At that time mine was the only show in town, and I knew that his threats were idle. He wanted the project to succeed as much as I did, and at no time was there any personal conflict between us. I truly admired him; ours was a conflict of style, only. I might add that he is currently the president of a large successful corporation where he has had an outstanding career.

8.6.2. Phosphate Fibers and Cement Pipes Are Incompatible

After one or two attempts failed, it was clear that there was vanishingly small hope that phosphate fibers could withstand pH 13, high-temperature aqueous solutions found within pipe, as manufactured in the United States. We made many attempts to protect the surface of the fibers in ways to help them through the curing process. If this had been successful, it is very likely that little of the safety factor built into the fibers would remain. All dissolved polyphosphates degrade rapidly, even in distilled water, at 170 °C.

8.7. Smokeless Cigarettes

One of the first requested potential uses of phosphate fibers was as the support body in smokeless cigarettes. There was no doubt that the product would have behaved well and would have been safe. Most of us associated with the product were not really too pleased that the fibers might be used in cigarettes of any kind. We were not displeased when the cigarette company decided not to market the cigarettes. To associate our product with yet another recognized hazard was not the kind of press we were seeking even if the product was safe and the effort noble.

8.8. Plastic Reinforcement: $[Ca(PO_3)_2]_n$ and $[NaCa(PO_3)_3]_n$

Despite the poor showing of calcium polyphosphate fibers in reinforcing Portland Cement, calcium polyphosphate and sodium calcium polyphosphate fibers were tested in plastics. In these tests $[Ca(PO_3)_2]_n$ fibers performed very well indeed. This was particularly true for epoxy resins.

Sodium calcium polyphosphate fibers performed similarly to "asbestos" in phenol formaldehyde resins. It did not compare well to potassium titanate in polyester resins. $[Ca(PO_3)_2]_n$ fibers, being developed by Abe in Japan, should perform much as potassium titanate, but without the hazards.

Reinforced Sulfur Pipe

Dr. John Hinkebein conceived of reinforcing elemental sulfur pipe with phosphate fibers. Elemental sulfur is used in a variety of sulfur cements and concretes and usually performs brilliantly, particularly under acidic conditions.[5] Plastic sulfur is strong when reinforced and withstands a variety of chemical and biological abuses. Dr. Hinkebein's method was to prepare pipe extrusion molded continuously in the field, where it was being laid continuously and at once. One merely heats sulfur above its melting point, 120 °C, while adding phosphate fibers and a compound to prevent sulfur from crystallizing. Pipe of almost any diameter can be continuously extruded, to great lengths, without the need for flanges or couplings. This concept has outstanding merit and should yield quality piping very inexpensively. The general appearance of this pipe is similar to PVC or ABS pipe, currently used extensively as a replacement for "asbestos"-cement pipe.

Although Dr. Hinkebein's concept was promoted as an application of phosphate fibers, it should be useful with other fibers. It is believed that, with some additional research, a very satisfactory replacement for "asbestos"-cement pipe can be obtained. The products made from this approach are an excellent candidate for replacement of both ABS and PVC products.

8.9. Phosphate Fiber Paper

The manufacture of phosphate fiber paper was a fascinating application of phosphate fiber. This test was semicommercial. A paper was obtained of very high quality and was further tested in several paper applications. Gaskets, ceiling covers, vinyl floor backings, and similar applications performed well. $[NaCa(PO_3)_3]_n$ fibers were used in these tests. The fibers behaved as expected throughout the run. Photographs of the equipment and resulting paper are seen in Figures 8.1 and 8.2.

Figure 8.1. Paper felt being made from $[NaCa(PO_3)_3]_n$ fibers.

Figure 8.2. Paper felt being made from $[NaCa(PO_3)_3]_n$ fibers.

Three weeks ago a gasket manufacturer called to tell me of their continued interest in phosphate fibers, because their gaskets tested in their long-term test had behaved excellently. "Were you considering manufacturing the fibers? I see so much written about the fibers." I had to tell him that we were not considering reopening our plant and that the articles he was reading were based upon a continued interest in the product because it had behaved well in most applications.

8.10. Automotive Brakes

In review, it becomes obvious that none of the rest of us associated with the Phosphate Fibers Project approached Dr. John A. Hinkebein in ingenuity, insight, productivity, and just pure genius. The application of phosphate fibers to automotive brakes is but one more example. Most of the data to follow were presented by Dr. Hinkebein to the Society of Automotive Engineers, 4th Annual Colloquium on Brakes, Atlantic City, New Jersey, October 20, 1986. Figure 8.3 is a photograph of automobile brake shoes made from phosphate fibers.

A review of the properties of phosphate fibers as related to the manufacture of brakes is presented in Table 8.2.

Dr. Hinkebein's conclusions were as follows:

Figure 8.3. Automobile brake pads made from phosphate fibers.

"The Phosphate Fiber contributes to processing and blending ease, uniformity of finished product, good high temperature friction, good resin absorption, and excellent cured strengths. It is not a plug-in replacement for short asbestos fiber, but can be formulated consistent with its own properties to produce excellent non-asbestos friction material candidates. Fine tuning of the formulation is dependent on other components of the formulation and to be specific processing conditions under control of the formulator.

Table 8.2. Fiber Properties Relative to Brakes

Aspect ratio measured	20 to 30
Effective aspect ratio	50 to 70
Average diameter	1 to 5 microns
Density	2.86 g/cm^3
Tensile strength	370,000 psi 2.6 GPa
Modulus	18×10^6 psi 124 GPa
Surface area	1,500 to 14,000 cm^2/g
Melting temperature	750 °C
Mohs hardness	3.7 to 4.3
Oil absorption	350 to 380 g oil/100 g fiber

It is hoped that the data given in this paper will be useful to friction materials formulators who desire to optimize the performance of Phosphate Fiber in their own formulations."

There is no doubt that these data should have been useful. Brake manufactures are still calling and requesting the fibers, while the EPA searches in vain for a replacement for semimetallic brake linings and is even considering returning to "asbestos." It has been recognized that more people are being killed by poor brakes than ever should have died from "asbestos." This should be no surprise to anyone.

In manufacturing processes for brakes, strength is critical in two different parts of the process. First of all, when the ingredients are mixed and pressed at 2650 psi for one minute in a cold mold, the friction block of molded ingredients must be strong enough to be handled when it is removed from the mold. Otherwise, the block will fall apart or crack before it can be placed into a hot mold for curing. The strength of the block as it is removed from the cold mold is called the *green strength* and satisfactory green strength ranges from about 20 to 90 psi as measured by flexural strengths.

It is obvious that the cured brake blocks must be very strong if they are to withstand the abuse that friction products must endure. The horsepower of an automobile may be as high as 50 to 100, but the brakes on the same automobile must be 400 to 600 horsepower if the brakes are to stop the automobile in one-sixth the distance required to accelerate the automobile, by its engine, to speeds where brakes are applied to stop the automobile.[6] The major fraction of the kinetic energy of the automobile is suddenly converted to heat in the brakes, and the temperature may rise abruptly from 0 °F to 1000 °F in a matter of seconds. Thermal shock alone would destroy an inferior brake pad.

8.11. Surgical Implants

Several groups have been studying both $[Ca(PO_3)_2]_n$ and $[NaCa(PO_3)_3]_n$ fibers for potential use as reinforcing fibers for bone implants and dental materials. Some of the results were encouraging, but naturally, much of the work was terminated when it was discovered that the product would not be available at any price.

On March 29, 1994, Paul Harvey, a well-known news reporter in the United States, reported that a child, a young girl, had been to the dentist. The dentist pulled one of her teeth. She requested that she be given the tooth to place under her pillow for the Tooth Fairy. The dentist informed her that he could not give her back her tooth because it was against OSHA regulations. I do not know whether the tooth was considered a toxic substance or a biohazard.

All phosphates are being considered as toxic substances. That includes your teeth and your bones. Maybe it is I who should be locked up to protect you from me. If it makes sense to class phosphates as toxic substances, there is surely something dangerously wrong with my thinking!

8.12. Miscellaneous Uses

1. Phosphate fibers should have been excellent materials for the insulation of pipes in submarines and similar naval uses. Its insulating properties are similar to chrysotile without the concerns that have been developed around that substance.

2. The fibers should have been an excellent additive for toothpaste to supplement the normal abrasives. They were soft and may well have behaved as a brush within the tube to assist both the toothbrush and paste without causing additional abrasiveness to a tooth.

3. Small fibers were used to flatten paint. They added texture to water-based paints and it was noted that the paint dried more rapidly when it contained a few percent phosphate fibers. This was a casual observation and more time would be required to understand why the paint dried more rapidly.

4. The fibers absorbed oil rapidly and should have been excellent fillers for oil filters. They should have had an added advantage of being a slow-acting desiccant in the filter and should have kept the oil anhydrous, thus eliminating the harmful influence of water in lubricating oils.

5. The fibers should also have been excellent filters for hydraulic lines where water can become a problem. The hydraulic brake of automobiles, particularly the master cylinders and wheel cylinders, are damaged by water collecting in the hygroscopic brake fluids. Nonhygroscopic brake fluids based upon silicones can eliminate this problem, but they are usually expensive.

6. The waste mash filtered from beer fermentation is often used as animal feed. If phosphate fibers were used as a filter medium, the filter could be mixed with the animal feed as a feed supplement. Calcium phosphates are regularly added to animal and chicken feeds.

7. Much "asbestos" was removed from hair driers and similar home equipment. The paper made from phosphate fibers should have been excellent replacements for these uses.

8. Air-conditioning ducts in homes and offices often had "asbestos" coverings for noise and fire control. Phosphate fiber paper should have been outstanding in these applications.

9. As mentioned under the section dealing with the electrical and static properties of phosphate fibers, they can impart antistatic properties to both plastic sheeting and cloth. No extensive work was done here other than to confirm that a static charge was controlled.

10. As an additive to hard surface cleaners, the soft fibers enhance the cleaning properties of existing cleaners and polishing agents.

11. Within the last week two new uses were called to my attention by potential customers seeking phosphate fibers for their particular applications. One was an automobile manufacture who had used the phosphate fibers in modeling clays while building dies for automobile parts. He stated that the phosphate fibers gave good

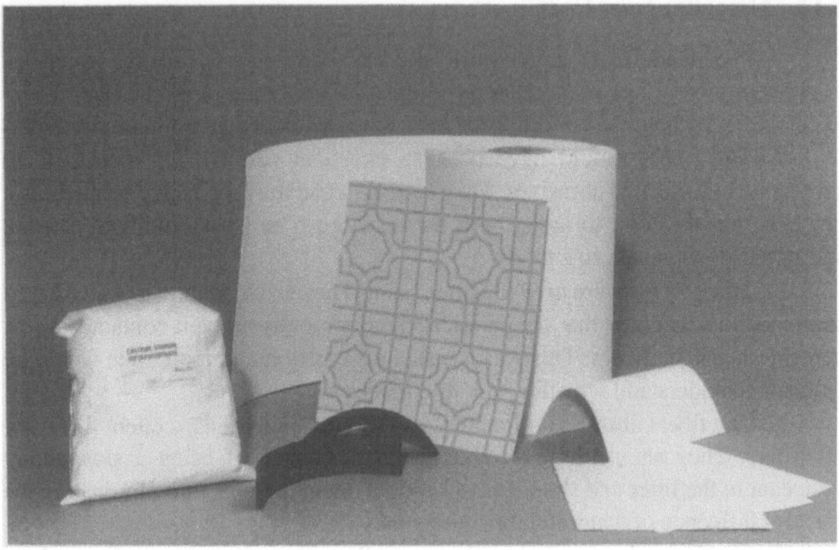

Figure 8.4. Products manufactured from phosphate fibers.

green strength and that the clays did not crack when dried or baked. The second person was seeking more phosphate fibers for a dental application for fillings. This would be a small marketplace, but could be supplied as a value-added item.

12. Before mineral fibers came under attack, chrysotile was used in electric cords to protect homes from fire damage. One of the problems that could arise was iron in the chrysotile. If it contained iron, the alternating current was inclined to cause the system to heat. It was necessary to remove iron from all mineral fibers that were used in electrical equipment. Phosphate fibers should have performed well in this application.

Many more items could be added to this list, but the general concepts being considered for the fibers are represented here. Figure 8.4 shows several products that were made from phosphate fibers in manufacturing facilities of several major companies that were interested in the product.

8.13. Another American Industry Is Gone Forever

Now that the phosphate household detergent industry has been completely disassembled and only a few derelict factories stand rusting and deserted, there are any number of articles being published with a message. "Phosphates are not as bad as we thought that they were." "Zeolites may be worse than phosphates." "Are Acrylics causing problems?" "Shell fish production has diminished now that the phosphates have been removed." "Phosphates are added to lakes to increase fish production." Three-year-old walleyes were reported to have come back to Lake Erie within *six months* of the time phosphate detergents were banned from the lake.

Having worked a lifetime with phosphates, I believe that phosphates are one of humanities greatest God-inspired gifts. They have been unfairly accused and maligned. They were first called nutrients, next they were labeled pollutants, and now of all misguided absurdities they are being named POISONS! Phosphoric acid is on the Toxic Substances List! Is nothing sacred?

8.14. Conclusions

Most of the more than 3000 known uses of serpentine minerals can be satisfied by the appropriate selection of phosphate fibers. Those uses where "asbestos" will withstand years of submersion in aquatic environments cannot be obtained with phosphates.

8.15. References

1. P. Gregory, in *High-Technology Applications of Organic Colorants* (A. R. Katritzky and G. J. Sabongi, eds.), Topics in Applied Chemistry, p. 176, Plenum Press, New York and London (1991).
2. J. Burke, "Enhancement of Alaska's commercial fisheries," *Alaska's Wild Life* **25**, 13 (1993).
3. L. A. Highfill, U.S. Patent No. 4,969,728 (October 16, 1990).
4. B. Z. Warren, U.S. Patent No. 4,988,522 (January 29, 1991); U.S. Patent No. 5,055,313 (October 8, 1991).
5. J. A. Hinkebein, U.S. Patent No. 4,484,950 (January 27, 1987).
6. *Hydraulic Disk and Drum Brake Manual*, p. 1, Wagner-Lockheed Brake Company, St. Louis, Missouri (1970).

9

Product Safety in the Climate that Phosphate Fibers Were Introduced: How Much Testing Is Enough?

9.1. The Author's Bias

It has been my steadfast conviction from the time I first conceived of phosphate fibers that they are a product that requires no toxicological study. This is the driving force that caused me to initiate the project in the first place. If the fibers are not obviously safe, it made no sense to begin a project. Fibers of phosphate are not new and have been used in all manner of household and industrial products for several lifetimes. They have never caused any toxic problems for people, animals, or vegetation. Literature mentions were made that many phosphates were fibrous and resemble chrysotile in appearance. No use had ever been made of their fiber properties. As long as they were merely being eaten, drunk, and inhaled, no notice was taken of the fact that they were fibrous. It was only when they were to become a safe replacement for "asbestos" that their safety was questioned. The same is true again today; we continue to eat, drink, and breathe phosphates and no one questions their safety.

Perhaps it was simple-minded of me, but I had not anticipated that the fiber's safety would be questioned by anyone. All types of phosphates have been commercial products for more than a century. No one has ever questioned this very same composition, when it is sold by another company as a potable water treating agent, Macromet®, to control rust in well water. It has been in service for many decades in drinking water. The phosphate fiber $[NaCa(PO_3)_3]_n$ is nothing more than crystalline Macromet®. Or conversely, Macromet® is phosphate fiber that has been melted to a glass. I had at one time considered purchasing Macromet® as a raw material from which to make $[NaCa(PO_3)_3]_n$.

Most water wells located in areas where the water contained much hardness and metals such as iron, lead, manganese, copper, etc. are treated with agents that are phosphate fibers or glasses for the protection of the plumbing, people, and animals using this water. Is this the type of product about which there should be reason to question its safety? Where is the *probable cause* that these products are harmful? No one has questioned fibers of insoluble sodium metaphosphate used in toothpaste and cheese. It would be a waste of effort because there is no *probable cause*. I accepted phosphate fibers as safe because in more than one hundred years there has been no probable cause that they are not safe.

Probable cause is a phrase well known in criminal law. It is unlawful to apprehend or search a citizen of the United States without probable cause that the citizen has committed a crime, felony, or misdemeanor. It is a part of our basic rights as citizens of the United States, along with being innocent until proved guilty. As stated earlier, no one is ever proved innocent. This is a reasonable course of action and precludes the requirement that all United States' citizens should be tried and purged routinely to prove that they are good citizens. In some other countries *probable cause* has not been required. Probable cause should be applied to substances as well as people. If it is not, there is a neverending wild attempt to prove the safety of the millions of compounds we all encounter every day of our lives.

There are some rules for substances that are intended to work like a *probable cause* for manufactured items. These are usually referred to as GRAS, generally regarded as safe. A simple formulation change may remove a substance from the GRAS list and the procedure starts all over again without probable cause. To start all over again is to initiate a new round of toxicity and other tests. Unfortunately, the tests are not definitive. When tests are completed we are still not certain whether or not a substance is toxic enough to be considered a threat to society. It will be seen that groups of conscientious professional scientists adopted diametrically opposite views concerning the safety of fibers, including phosphate fibers. Now there is probable cause to question the safety of the fibers, *but no way known to humanity to prove a substance is safe*. It is impossible.

9.1.1. Animal Tests

Some of the test methods used in toxicology are never encountered except in laboratory environments. However, much of our protection from toxic substances and progress in fighting diseases has been a direct result of tests with animals. If we are to continue to learn to protect the health of men, women, children, and animals, we have no choice; we must continue to learn from animals. This is not a point of contention. Toxicity tests may at times test animals rather than substances.

This is particularly true when tests expose animals to unrealistic conditions even when the tests are in no way harmful to the animals.

Most children know that wood splinters can cause serious problems when left embedded in a toe or finger. If not removed, splinters can cause infection and tumors may also occur. There are many examples of persons dying from tetanus caused by infected splinters. If a person's abdomen were opened and stuffed with wood splinters of calibrated sizes, this might serve as a test of hazards of wood splinters. It could be used to determine whether or not there is a difference between pine, ash, willow, and oak splinters when we imbed them in our chests between our lungs and pleural cavities.

In some cases tumors are certain to be caused by implanted wood splinters. Suppose a Splinter Crisis were well publicized by the media and a high official was to publicly proclaim a state of emergency: "Scientific evidence implicates wood splinters as a cause of CANCERS in test animals. When splinters are implanted in the pleural cavity of tumor-prone male kangaroo rats that cannot jump fast enough to avoid capture, a statistically significant number develop uterine cancers." Wood splinters could join "asbestos" and no more wooden homes could be built. All trees would be stenciled in pink ink, "Trees may contain splinters that are hazardous to your health." "Wood splinters have been shown to be fibers similar to 'asbestos'."

Your first reaction to a Splinter Crisis is probably, "Ridiculous!"

A moment's reflection will reveal that several items of commerce have been treated in this way and a Splinter Crisis is not as far fetched as it might at first seem. Several examples will be noted. I felt that the attack on phosphate detergents was a Splinter Crisis. I could not believe what was happening and it is difficult for me to accept it happened, even today.

To subject sodium or calcium polyphosphates to extensive toxicological studies, after millions of years of ATP in Nature's laboratory, seems to be an exercise in the obvious, similar to the Splinter Crisis. This precludes, of course, opening an animal's abdomen and injecting quantities of the foreign substance. When faced down by a roomful of disapproving toxicologists, looking over and under their eyeglasses at me, I reluctantly submitted phosphate fibers for implant tests, but never would I have agreed to gelatin–phosphate fiber implants. The tests I agreed to were free fibers in the peritoneal cavity. I did not like the tests, but it was obvious to me that I would never receive the approval of the medical community if phosphate fibers were not subjected to extensive rigorous tests.

"It is protocol," whatever that means.

Fibers of any kind are another issue when dealing with toxicology. As mentioned in Chapter 1, all fibers including cotton candy are potentially harmful in some manner. If a substance is fibrous, it is automatically considered to be a likely source of problems, unless there are no problems that can be identified. If one group finds that a substance is definitely a carcinogen and another group of equal rank can find nothing wrong with the fiber, it is likely that those predicting hazards will dominate, even after it has been shown that their results were in error. It is the safest course of action for everyone and would be my personal reaction.

A new product usually dies in its tracks when attacked by almost anyone for almost any reason. Products that are truly new are very fragile under the best of conditions and are very difficult to defend. Most new products never make it to the marketplace with or without safety constraints. There is no way to judge the value of a product to society or to a company before it has been accepted or rejected in a marketplace. Even this may not be a definitive test, particularly if the timing of introducing the product is poor. Management and sales people like to threaten research inventors with a "window of opportunity." A window of opportunity truly exists, but it is usually much broader than those pushing a project will admit. The width of the window usually depends upon the value of the product to the public.

Many products that ultimately proved to be indispensable were rejected before the public had a chance to test their true value. The Xerox® copying machine has been mentioned. The Post-it® notes of the 3M company were rejected by the company management before the secretaries of 3M all but demanded more of the wonderful little note pads. This list could continue for many pages, but illustrates the difficulties one ill-conceived toxicity test can cause a potentially useful new product that must overcome all manner of obstacles, irrespective of its safety. If Necessity is the Mother of Invention, Invention is the Grandmother of Necessity.

Later in this chapter, several references will be cited where groups of fibers were tested by Adduce *et al.* These studies found several fibers to be suspect, including phosphate fibers. This research showed no evidence or probable cause that rock wool, potassium titanate, or crocidolite might pose a health hazard, while amosite ranked among the less harmful! Chrysotile, the "asbestos" of the Americas, was not tested. These results open a Pandora's Box for me. I have no choice other than to report the results to you, even though I doubt they are meaningful. These results disagree with too much previously accepted work and the results we developed in our toxicity tests. However, as pointed out several times, the mere fact that results have been widely accepted does not necessarily mean that they are correct.

Results of an involved, long-term, carefully executed, inhalation study revealed no significant problems caused by phosphate fibers. I am reluctant to accept the hazardous results of the Adduce group. If phosphate fibers are truly hazardous, I do not wish to "whitewash" them, while making believe that they are safe when indeed they might be hazardous. But, if they are safe, as I believe them to be, I do not wish to indict them in a way that might ultimately deprive humanity of free access to them. It is my heartfelt belief that phosphate fibers will yet serve humanity in the ways I first envisioned.

Throughout this book I have stressed that our company chose not to manufacture phosphate fibers, not because they were believed to be hazardous, but because phosphate fibers were perfect targets for lawsuits. This remains our stance on this issue. Literature reports such as those cited may or may not be accurate, but are certain to cause difficulties in manufacturing and selling a new fiber regardless of how safe it may actually be. These approaches to safety are causing research on new materials to slowly grind to a halt. It is necessary that a compromise between risk, environment, and utility be addressed more realistically than has yet been done.

Unfortunately, decisions on the ultimate hazards of a product are not determined by either toxicologists or scientists, but by attorneys and jurors in a court of law where no one is likely to have a clear understanding of words such as "physical chemist." What does a physical chemist really do in an industry? I ask the question because I am one and I have practiced the science for almost half a century. It is also unfortunate that most industrial management does not have a foggy notion of what a physical chemist does. Most other types of chemists are not too comfortable with their definitions.

The financial risk is too great to introduce a new product that may or may not be accepted by customers only to be destroyed by regulations. One can never be certain that customers will like a new product. An excellent example was the leather replacement, Corfam®, introduced by duPont. Everyone knew it was magnificent, except the customers. I thought it was magnificent and still wear shoes made from Corfam®. Corfam® failed even before regulations entered the picture. Today, a huge sum of money can be spent on research and regulation clearance, only to suffer lawsuits that may or may not be groundless. The plaintiff usually wins some kind of a settlement in these cases. These facts cause the bravest of managements to become deliberate and shy. The number of truly new substances being manufactured is becoming alarmingly small.

In today's world, even the obvious is not obvious, unless it has been absolved by a ritual on an altar erected to some abstract gods of purity. It is as though a substance must be baptized and sanctified in some ceremony, to drive away evil spirits, before it can be approved for general use. In the words of the late Dixie Lee Ray, "What ever became of common sense?"[1] There are definitely substances and processes that require study by an independent group before they are put in a market for sale. These are usually obvious, but does this mean that everything must be tested and retested for each new use? I have been involved, as a consultant, in obtaining a clearance to put a fertilizer on the ground. This fertilizer is a potassium phosphate fertilizer that has been used as a fertilizer for more than 100 years. This fertilizer is used directly in foods to be eaten by people and is cleared for this use.

Surprise! Surprise! The vintage fertilizer was cleared to be put on the ground, but only after large sums of money were spent to have it cleared. We must question whether or not safety is the real motive behind the clearance of a product when the chances that the product could be unsafe is almost zero even before it is tested.

9.1.2. Saccharine, Dioxin, and NTA

Phosphate fibers were to be a product of a company that at one time in the past had manufactured saccharine, NTA, and traces of dioxin as a byproduct in Agent Orange. These three products are background and help to explain some decisions made by management in the Phosphate Fibers Project. We were also a manufacturer of phosphates for detergents. You can understand that our management had but small choice to move very slowly and deliberately with the Phosphate Fibers Project irrespective of the benefit to humanity. History had taught us that even the noblest of efforts can be turned sour for the providers.

Many examples of inconsistencies in our present system come to mind. Consider the cases of saccharine, NTA, and dioxin. Since most readers are no longer familiar with NTA, perhaps it is a good starting material upon which to initiate this comparison. NTA is shorthand for nitrilotriacetate. It is a simple compound that was manufactured as a replacement for detergent phosphates, when detergent phosphates were beginning to be banned in the United States. NTA is still used in detergents in Canada, as it has been for three decades.

It has now been many years since all active substances in detergents were condemned one right after the other. First it was alkyl benzene sulfonates because they foamed. This was followed by sodium tripolyphosphate because it is a nutrient, and then NTA as a result of ignorance and bad data. Within ten years of their introduction the finest detergents ever manufactured were under mortal attack. Before the phosphate detergents there were no satisfactory automatic washing machines.

Without phosphate detergents there could be no automatic washing machines. Soaps and washing powders simply will not work. Most women spent Mondays doing the family's laundry. Phosphate detergents as much as any other single item freed American women from Monday washday blues. At last a woman could take her rightful place in society and business without being burdened to hand-wash clothes two days each week.

We scrapped phosphate detergents by mandate. Billions of dollars of wasted effort later, no detergent builders have been found that approach phosphates or NTA as detergent builders. NTA was surely the best substitute for phosphates ever found, but even it was a pallid second to polyphosphates, when "cleanliness is truly next to godliness and good health."

Hundreds of scientists with the highest credentials in many companies throughout the world had worked on NTA for several years before it was introduced as a new detergent builder. Like phosphates nothing could be found wrong with the compound. Both our environment and the general public should be served well and safely.

Despite what some misled persons seem to believe, industrial scientists are among the more dedicated professionals anywhere in protecting our environment and the customers they serve. A manufacturing plant had been built to supply NTA. Detergents of high quality were being accepted by the public. I was not too happy that NTA was as good as it was because I was promoting phosphates for detergents. Nevertheless, I had to admit NTA is a good, safe detergent builder.

BAMB!

Out of nowhere the NTA industry is attacked. Two technicians working in a government laboratory made a remarkable, profound, unexpected discovery.

9.1.3. A Teratogen?

NTA is a teratogen!!!!

"NTA is a what?"

I didn't know what a teratogen is. I was not the only person scampering to a dictionary for a definition.

The best that I can remember how the NTA episode was relayed to me is that the story went something like this. If a pregnant female, who was approximately six weeks pregnant, drank water from an *anaerobic* well that contained NTA, and cadmium, or mercury, her child was likely to be deformed! (If a pregnant woman drank water from a well that contained significant quantities of cadmium and/or mercury, it made small difference whether or not the well also contained NTA.) These terrifying results were reported by a public health official. He publicly requested in a radio and television broadcast that industry immediately discontinue the manufacture of NTA for detergents. It is a crisis. This may not be exactly as it was. It is surely the gist if not the exact details.

The government laboratory results were wrong. Right or wrong, the results killed the product in the United States forever. Reports that a product is harmful is easily believed, while reports that a product is harmless are usually ignored or never written about, but not always. Can you imagine a high-ranking government official requesting time to tell you that it is safe to eat ice cream or paint with white paint? It will be seen that another official, from the same office as the one who killed the NTA industry with one announcement, was ignored when a similar announcement was made about saccharine. NTA was announced as though it were a crisis requiring immediate action, while the saccharine story evolved.

Industry had no choice other than to discontinue manufacturing NTA, for use in the United States. This one announcement murdered the product for use in the United States, forever, but not in Canada where it is still used without deforming Canadians. It has always been a bit baffling to me why the Canadian people are so very different from those of us who live south of their border. Canadians can use NTA with no problems, but we must not use NTA because it will deform us. Perhaps the closer to the equator one moves, the more toxic NTA becomes.

That's toxicology.

Or is it?

You can be certain that there is something about NTA that is harmful. I don't know what it is, but there is surely something wrong with it. There is something harmful with everything that is functional. If we look hard enough and long enough, it is merely a matter of time before we know what it is. Lucretius recognized this more than fifty years before the birth of Christ,

"What is food to one, is to others bitter poison."[2]

I did a little work on the NTA project, but I was not a principal investigator. When careful, comprehensive research was conducted, it was shown that NTA was not a teratogen, but actually protected a fetus from harm by water that contained a few parts per million of most heavy metals. No one requested air time to tell you, "Sorry we were wrong. Go back and build confidence in your customers." The *post mortem* was meaningless, because the product was gone. No detergent manufacturers in their right minds would attempt to sell a detergent to a public, in contradiction to the popular notion, spawned by leading officials of our government and media. Where are responsible manufacturers who would attempt to sell a product, while the product had the remotest possibility of harming unborn children? Lawsuits would flood the company for every deformity of a child, even before the first boxes settled on the grocer's shelves.

A second case study is the well-known saccharine story. An official announced that saccharine is a carcinogen. The public answer was:

"Who cares!"

"Caveat Emptor!"

"Write it on the back of the package!"

"Let the written word scare the evil spirits away!"

Now we are protected by *the printed word*, on the back of every pink package, telling us that the Surgeon General is a believer. "*Use of this product may be hazardous to your health. This product contains saccharine which has been determined to cause cancers in laboratory animals.*" I have never been certain what these words from on high were intended to accomplish. They seem to stop no one from eating saccharine. Judging from the number of pink packages to be found on restaurant tables throughout the world, it seemed to have increased the sale of saccharine. But a similar approach to smoking has surely done much to save lives. It is difficult to predict how we will react to information.

Back to sweeteners; far less damning evidence completely destroyed calcium cyclamates. There were rumors in the scientific community that the data killing cyclamates were intentionally designed to kill them. This may or may not be true, but in view of some other things happening at that time, I am inclined to believe that it is true.

The political poison, dioxin, ranks high on the list of trumped-up charges. If Agent Orange had not contained traces of dioxin, few people would ever have known there are such compounds. Some leading household and office products had contained traces of dioxin for many years with no obvious harm to the users. As mentioned so many times in this book, there is no way known to human logic to prove that a substance is not harmful. As a result of this, the fact that little or nothing can be found wrong with a product does not vindicate those who have some ulterior motive to attempt to prove that it is harmful. There is little doubt that those who wish to keep dioxin a political poison will continue to do so long after they have forgotten why they thought it was harmful in the first place.

Unfortunately, phosphates fibers are a fourth substance that should be added to this list. They suffered a similar, but less publicized fate. The belabored examples given above were chosen to better illustrate the climate under which phosphate fibers were invented, researched, and manufactured. No new product could be prepared without being challenged, and particularly one that was to be a substitute for a substance that was used to frighten the public of the many hazards they were facing.

9.2. Are Phosphate Fibers Carcinogens?

I had hardly demonstrated that useful sodium and calcium phosphate fibers could be made before their safety was questioned. The first to question their safety was a business school type, vice president. He was in the audience when I made the first public disclosure of phosphate fibers within our company.

The Vice President approached me, the one who ultimately controlled our research money, and asked "How do you know they don't cause cancer?"

I was insulted. I answered, "If they are unsafe, they are the first sodium or calcium polyphosphate to be unsafe."

He countered, "That's what *THEY* said about asbestos."

The VP knew nothing about "asbestos" and less about polyphosphates, but business schools had told him that he "didn't have to know nothin about nothin, to manage everthing." He believed it.

I do not!

One of my scientific colleagues summed it up when he quietly approached me after the talk when the VP had left the room, "You cast pearls before swine!"

I have small doubt that experiments can be contrived that will give positive results that phosphate fibers are carcinogens. I believe that it is safe to state that any substance found on Earth is a carcinogen, properly applied with the intentions of creating cancers.

Our data support that when used with normal precautions phosphate fibers are NOT carcinogens.

Yes! I remain bitter on these points!

9.3. Phosphate Fibers on Trial

From the time of our very first phosphate fiber safety meeting, it was obvious that phosphate fibers and I were on trial. We were to spend more time and money on toxicology than we were to spend on research to make fibers useful enough to demand a market. No matter how safe a substance may be, if it has no use, the safety question is moot. On the other hand, if a substance is functional, if it does any useful function, it will be viewed as both good and harmful. If it is useful enough to be desired by large numbers of persons, there is no doubt that it will be labeled harmful, if for no other reason than it takes up space and there will be a lot of it in our environment. A moment's reflection will point up the validity of this observation.

Many products are sold today with a label listing what is *not* in the box. A recycled green box that contained absolutely nothing should sell well. It requires almost no knowledge to cite problems with just about anything useful. Unfortunately, those who can find problems seldom know how to solve the problems they discover or know whether or not they are truly problems. With my great respect for, and agreement with Voltaire,[3] "this is not the best of all possible worlds," nor are we the brightest of all possible peoples.

In the early days of the Phosphate Fibers Project it was my responsibility to hold the first safety meetings to discuss how we should test phosphates. While we discussed this problem, I served coffee and phosphates in the bakery goods eaten by the group. I made a point of mentioning that they had just eaten some of the substances we were to be testing.

The initial probe was to test how rapidly a phosphate fiber would disappear if imbedded in a living organism. These groups seem to be obsessed with imbedding something in something. I already had a reasonably good idea of how long a phosphate fiber should remain in an organism.

The First Test of Phosphate Fiber's Safety

Some early calcium polyphosphate fiber batches contained long sharp needles. These needles were perhaps one inch in length and maybe two millimeters in diameter, but sharp on the ends. Quite by mistake, I impaled my index finger on one of these needles and it broke just below skin level. My first thought was to excise the fragment. Then it occurred to me that this should be an ideal time to test how long the bit of fiber would remain in my finger. I had cut my hands many times with bits of phosphate glass and none had ever caused any trouble.

The skin over the small shard healed with no lump, infection, or other trouble, except that when I would shake hands with someone, pressure on the fragment produced some mild pain. I recorded the incident in my notebook while keeping a record of how long I could continue to feel the fiber in my finger. Within three weeks there was no sign that my finger had been injured. This was the first implant test of a phosphate fiber.

9.4. Microfiber, $Ca_2P_2O_7$ Implants

In an attempt to grow microfibers, some work was done with calcium pyrophosphate. It was similar to the calcium pyrophosphate that had been used for many years as the abrasive in the world's leading toothpaste. Its morphology had been changed slightly. An experiment was planned to place this calcium pyrophosphate microfiber in an atomic pile to activate it, and then to implant these fibers in peritoneals of rats. Movement of this phosphate could be monitored in several ways. Urine and feces from these rats were analyzed for activity, while radiographs were made of the whole rats. Within six months it was obvious that the phosphate would not remain in the peritoneal of these rats. The phosphates had moved throughout the rats and much had collected in the teeth and bones of the rats. These phosphates behaved exactly as they were expected to behave. This is what happened in my finger; the phosphates were converted from fibers to nutrients.

9.5. Professional Tests

The safety tests done, in house, on phosphate fibers were conducted by a group of highly skilled, dedicated scientists. They remained objective throughout the test. I felt that they were overly cautious, but it is much better to be cautious than complacent when so much depends upon the results of the tests. All members of the Phosphate Fibers Project group are deeply indebted to Dr. Rashmi Nair and her fellow workers for the care they exercised to be certain that phosphate fibers were not unsafe and could be used with confidence.

I disagreed with but one test done by our toxicologists. In Chapter 1 there is mention of a gelatin pledget test. Phosphate fibers did not do poorly, but I believe the test is meaningless. Chemistry is based upon a simple concept. When two or more substances react, the products of the reaction are new substances unlike the reacting substances. The reaction of polyphosphates with proteins is not exactly a new revelation. J. J. Berzelius[4] published this information in 1816. When allowed to become moist and to react with gelatin, polyphosphate imbedded in gelatin is neither polyphosphate nor gelatin, but becomes an insoluble coacervate of the two.

The test violated two of the governing objectives for preparing phosphate fibers. It immobilized the phosphates (Focal Point 5). The first animal test was to examine phosphate fiber's mobility in living systems. A part of the harmful influence of "asbestos" is believed to be immobilization. Fibers cannot be flushed from the body. The serpentine fibers are so inert chemically that they cannot degrade or dissolve. In the test, gelatin reacted with any polyphosphate that attempted to dissolve, preventing its degradation and violating the second governing guiding concept that promotes phosphate fiber's expulsion from living systems (Focal Point 8). Some phosphate fibers and gelatin were still in place when the rats were sacrificed two years later.

The gelatin test was performed by a so-called Stanton protocol, taking a small 0.2 cm × 0.3 cm square of dry gelatin foam and imbedding in it 40 mg of polyphosphate. One-half ml of a ten percent gelatin solution was then poured over the fibers to seal them in place. The gelatin pledget was next placed in the pleural cavity of a rat between the lung and the cavity wall. This is not a very moist area. The rat breathed against this thorny cocklebur for two years. Considering the nature of the test, it is surprising to a chemist that only three rats in two years developed lung tumors. If this is a typical test that toxicologists use to judge the potential hazards of solids, no wonder so many substances are believed to be dangerous to our health.

In the gelatin test, it should not have made much difference which polyphosphate is chosen, from toothpaste, to leavening agent, to water softeners, to phosphate fibers. All polyphosphates would have tanned the gelatin to a leather-like solid as the phosphates hydrolyzed and became acidic. In this case the gelatin would not have dissolved any more than it did in the rat test. The idea was, of course, that the gelatin was to dissolve leaving the free fibers. With an inert fiber that did not react with the gelatin this may happen, but not in the case of polyphosphates and gelatin.

9.6. Toxicology Studies

In a more recent fiber study in Japan and mentioned above, synthetic fibers were compared with mineral fibers. Nine fibers were tested.[5–7] In each study 20 hamsters were chosen for each fiber tested for a total of 180 animals plus controls. The fibers were characterized by scanning electron microscopy, and the pictures were of excellent quality.

Samples of each fiber were prepared by suspending 500 mg of fiber in 50 ml of saline solution, to which had been added 250 mg of carboxymethylcellulose to prevent the fibers from settling. These suspensions were injected into the trachea of hamsters. Once each week for five weeks, 0.2 ml suspension and 0.3 ml of air was injected into the hamsters. The hamsters were sacrificed over a two-year period.

Some very questionable results were obtained. One of the more interesting results was the fact that tumors were not confined to the pleural cavity, but developed at sites far removed from the trachea and lungs in many cases. Three substances tested that have a long history of causing tumors in animals were amosite, crocidolite, and potassium titanate. The amphibole crocidolite and potassium titanate showed no tumors, while amosite and fiber glass ranked as the least suspect! The results were as presented below. They contradict almost all studies published and the results with respect to phosphate fibers cannot be accepted as reliable!

	Tumors/ 20 Hamsters
1. Carbon fibers	0
2. Potassium titanate	0
3. Crocidolite	0
4. Rock wool (this requires definition)	0
5. Amosite	2
6. Fiber glass (this requires definition)	2
7. Calcium sulfate (Franklin fibers)	3
8. Phosphate fibers, $[NaCa(PO_3)_3]_n$	5
9. Basic magnesium sulfate	9

Everything else being equal, this study suggests that the more quickly a substance dissolves, the more likely it is to form tumors when injected into the trachea of hamsters. Basic magnesium sulfate is usually represented as $6(Mg(OH)_2 \cdot MgSO_4 \cdot 3H_2O$ and is reported to be sparingly soluble.[8] It is to be expected that the solution of the salt might be basic, but it is not obvious why this compound should have a 45% tumor-causing rate when both $MgSO_4$ and $Mg(OH)_2$ have been ingested and inhaled for years without any significant problems being reported. Aqueous suspensions of $Mg(OH)_2$ are known as milk of magnesia, while $MgSO_4 \cdot 7H_2O$ is Epsom Salts. The recommended dosage of Epsom Salts for horses and cows is four pounds each.[9] From these results one should expect table salt to be an arch killer! It should be tested by this technique and with hamsters.

The question again arises, "Are we testing substances or animals?" In the study discussed below, rats were tested rather than hamsters. The results will be completely contradictory. Are both groups correct and hamsters and rats cannot be compared? Do animal tests have any meaning with respect to human subjects when fibers are being tested? There seems to be no answer to these questions. In the work with dioxin, there was no doubt that guinea pigs were highly susceptible to dioxin poisoning while nothing else seemed to be.

Not included in the fiber study above were: (1) chrysotile, (2) phosphate fibers, $[Ca(PO_3)_2]_n$.

Several toxicologists were consulted as to their opinions concerning tests in which fibers are suspended in media and then injected into tracheae of animals. Their primary concern was different from my own. Two different groups said that fibers are inclined to form clumps in the lungs. These lumps are not characteristic of air-borne fibers. My concern was that fibers were washed into places they might never have reached if they had been air-borne. In the work to follow some criticism was made of our work with rats, because rats might have an electrical charged condition in the respiratory system that could trap air-borne fibers, keeping them from the lungs.

In the Monsanto study of phosphate fibers only, $[NaCa(PO_3)_3]_n$ was intensely investigated with air-borne fibers, but samples were also injected into the tracheae of rats in much the same way they were in the Adduce study. It was felt that air-borne samples were a more realistic test than attempting to inject fiber samples into the animals to study the potential hazards. It will be seen that even when the samples were injected, the results differed from the Adduce study. It was observed that fibers did form lumps when injected in aqueous fluids.

Phase I of the toxicological study was undertaken to determine whether or not some of our initial assumptions (Focal Points) were correct in biological systems. Phosphate fibers were studied in biological media similar to those encountered in living organisms.[10] Rates of solution, hydrolytic degradation, and relative toxicity were determined in some very elegant and definitive experiments. The physical chemical studies were probably some of the more reliable ever reported. Fortunately there were no surprises.

Monsanto Intensive Studies

Phase I

Perhaps there is no better way to introduce Phase I studies than to quote directly from the article, *Toxicity Of Calcium Sodium Metaphosphate Fiber, 1. In Vitro and in Vivo Degradation and Clearance Studies.*[11]

> "In order to test our hypotheses of biodegradability and the contribution of this property to ultimate toxicity of the fiber, a two-phase testing program was designed. In Phase I, the degradability of the CSM ($[NaCa(PO_3)_3]_n$) fiber in vitro and vivo and clearance from the lung was evaluated. In Phase II, the long-term toxicity of CSM ($[NaCa(PO_3)_3]_n$) fiber was evaluated (Nair et al., 1992). Phase I existed of three experiments. Experiment I was carried out to determine degradation of CSM ($[NaCa(PO_3)_3]_n$) fiber in the presence of cultured mammalian cells. The cells used were lung epithelial cells (LEC) and alveolar macrophage of rat origin (RAM). These cells represent the major cell types that may have direct contact with the fibers upon inhalation exposure. Experiment II was carried out to determine in vivo degradation of the fiber as followed by whole-body autoradiography. The purpose of Experiment III was to determine the clearance of CSM ($[NaCa(PO_3)_3]_n$) fiber from the lung after intratracheal inhalation (IH) and intratracheal instillation (IT). This paper discusses the results of Phase I of the testing program."

In the Phase I studies it was shown that in cell cultures $[NaCa(PO_3)_3]_n$ fibers degraded and cleared as predicted. Rather than attempting to give a detailed analysis of the work here the reader is referred to the original article.[11] There, the minute compliances with the Good Laboratory Practice Regulations dictated by the EPA are related. It will suffice for our purposes to quote the summary statement from the article.

"In summary, all of the studies discussed show that CSM ($[NaCa(PO_3)_3]_n$) fiber is biologically degradable and at the concentrations studied in intratracheal instillation and inhalation studies produce a minimal histological response. The toxicity of the fibrous material appears to be directly related to the durability and dimensions of the fiber (Stanton et al., 1977, 1981; Kuschner, 1987). Thus the present study suggests that CSM ($[NaCa(PO_3)_3]_n$) fiber should have a low chronic toxicity potential since the fiber appeared to be nondurable in biological systems."

The above conclusion is general, but is subject to the limitation surrounding most biological systems. It is almost impossible to make a definitive statement. This is *not* intended as harsh criticism of this or any other toxicity work, but is merely a continuation of the theme that it is almost impossible to judge the impact of a chemical from laboratory work or short-term case studies. The systems are very complex and all variables cannot even be identified, much less controlled. The summary is written to anticipate studies such as the Adduce et al. work. When diametrically opposite conclusions are drawn about a system studied by two independent groups, unknown variables must be influencing the results. Perhaps both groups are correct under the design of their experiments.

Phase II

Phase II of the testing program with CSM ($[NaCa(PO_3)_3]_n$) consisted of a chronic inhalation study. One hundred sixty rats were used in the test. They were divided into four groups and exposed to air-borne fibers of 0, 1, 5, or 25 mg/m^3, six hours per day, five days per week for two years or until the rats were sacrificed at predetermined intervals. The details of the study can be found in the published data.[12]

The conclusion of the study was as follows:

"The latter {$[NaCa(PO_3)_3]_n$ fibers} shows only a minimal potential at the high-exposure level and the absence of a tumor induction at any of the exposure levels. In addition, it is significant that there was no pleural reaction to the fiber either in the present chronic inhalation study or in the 6-month intratracheal instillation and intratracheal inhalation study. In conclusion, CSM ($[NaCa(PO_3)_3]_n$) fiber should have low potential for producing chronic adverse health effects, i.e., fibrosis or neoplasia, following inhalation exposure."

9.7. Silicate–Phosphate Toxicity

Abe's group uncovered the disturbing bit of information about the toxicity of phosphate–silicate reacted composites and is worthy of being mentioned.[13] The fault seems to be a result of silicate–phosphate mixed SiOP linkages rather than silicates or phosphates alone. But it does suggest that when working with chemically bonded mixed anions, care must be exercised. It also suggests that when working with amorphous phosphates containing heteropoly anionic linkages of any system of phosphates, care should be exercised that one does not become too complacent. At this time there is no reason to believe that unreacted mixtures of anions should cause problems.

9.8. Superoxide Forms

When composite glasses of silicate–phosphates are implanted in the bones of mice, they become toxic to the mice. It is believed that there is a production of superoxide during the hydrolytic degradation of a POSi linkage. It is the oxygen that is toxic to the mice, but nevertheless it raises a flag of concern if these MIXED HETEROPOLY MOLECULAR composites are considered for either dental or orthopedic work.

It must be emphasized that neither phosphates nor silicates are implicated, nor are admixtures of the compounds. It is only when POSi linkages are intentionally formed in single molecules that problems were noted. This information is sufficient to incite the ignorant, supersafe, aggressive protectors of all generations to come, to mount yet another crusade, to save us all from self-destruction! Their message will probably be, "Earth is composed of tons and tons of both phosphates and SiO_2. We cannot escape them! Surely, generations yet to be conceived are doomed to die!" Hopefully we can stop this one before our saviors have their first rally.

Samples of calcium silicate–phosphate glasses were prepared with various ratios of SiO_2 to P_2O_5 ranging from pure SiO_2 to pure calcium phosphates. The glasses were ground to powders of approximately 5 microns, suspended in saline solutions followed by intraperitoneal injections into mice. The compositions containing 30% SiO_2 were particularly harmful, yielding 100% mortality within 72 hours. Although the exact mechanism is not completely clear, it is believed that the superoxide is generated during the hydrolytic degradation of the SiOP linkages, because the solubility of SiO_2 reaches a maximum near 30% SiO_2 concentrations with respect to P_2O_5. The confirmation of superoxide was done with ferricytochrome reduction.

The results of the above work should surely be accepted as a warning sign to any groups considering silicate–phosphate compositions for either bone implants or dental applications. But it should also be re-emphasized that this is *not* applicable to either silicates or phosphates that have not been intentionally formed into glasses at very high temperatures! The spun fibers of amorphous phosphate used in bone implants, mentioned in Chapter 6, were reported to be modified with SiO_2 and other minor additives in some cases.[14] Although these fibers were reported to show no toxicity in implants, the fibers containing SiO_2 are at least suspect and should be utilized with care.

One other point should be mentioned before leaving the subject of silicate–phosphate glass toxicity. There is no reason to believe that if these glasses were either inhaled or ingested they should be toxic. It is only in the case of subcutaneous implants that superoxide could form in toxic quantities.

9.9. Conclusion

Phosphate fibers have been grouped with many other substances that have been rejected for many reasons. Case histories have been included of substances that were banned for lack of a method to prove that they are safe. There is no way known to prove a substance is safe and it is doubtful that there shall ever be. Our only choice seems to be that we must protect ourselves from those items that are recognized as harmful while respecting the potential for harm in all substances. We cannot and we must not attempt to isolate ourselves from everything we do not understand because our understanding of every substance, even when isolated from all other substances, is extremely limited.

As I completed this chapter I was reading the Hon. Stephen Breyer's book, *Breaking the Vicious Circle.*[15] Judge Breyer is the newest judge of the United States Supreme Court. I was truly amazed at the judge's uncanny insight and comprehension of the problem facing not only the peoples of the United States, but of the rest of the world. The rest of the world has followed our lead. *Breaking the Vicious Circle* should be required reading for every member of Congress and I strongly recommend it to you. This book, more than any other I have read, relays the message I have been attempting to bring you. *Breaking the Vicious Circle* says it much better than I did. I consider it the best work I have ever read on this subject.

9.10. References

1. D. L. Ray, *Environmental Overkill, Whatever Happened to Common Sense?*, Regnery Gateway, Washington, D.C. (1993).
2. T. Lucretius, [Titus Lucretius Carus] 99 to 55 BC, *De Rerun Natura*, Vol. IV, line 637.
3. F. M. A. Voltaire, *Candide* (1759).
4. J. J. Berzelius, *Ann. Physik.* **53**, 393 (1816); *Ann. Physik.* **54**, 31 (1816); *Ann. Chim. Phys.* **2**, 151, 217, 329 (1816).
5. S. Adduce, K. Kawamura, S. Yoshida, and K. Takemoto, *Int. Arch. Occup. Environ. Health* **63**(8), 553 (1992).
6. S. Adduce, K. Kawamura, K. Kimura, and K. Takemoto, *Environ. Res.* **58**(1), 55 (1992).
7. S. Adduce, K. Takemoto, and K. Kimura, *Environ. Res.* **54**(1), 52 (1991).
8. P. Brodeur, *Expendable Americans*, Viking Press, New York (1974).
9. J. W. Mellor, *Inorganic and Theoretical Chemistry*, Vol. IV, p. 332, Longmore and Green Co., New York and London (1942).
10. P. G. Stecher (ed.), *The Merck Index*, 8th edition, p. 638, Merck & Co. Inc. Rahway, N.J. (1968).
11. R. S. Nair, A. P. Li, R. N. Shiotsuka, and A. G. E. Wilson, *Fundam. Appl. Toxicol.* **19**, 68 (1992).
12. R. S. Nair, F. R. Johannsen, H. F. Bolte, P. E. Newton, and W. E. Rinehart, *Fundam. Appl. Toxicol.* **19**, 79 (1992).
13. Y. Abe, H. Hosono, T. Kasuga, and M. Nagase, *Phosphorus Sulfur Silicon* **76**, 243 (1993).
14. Y. B. Peng and J. B. Koen, *Bioabsorbable Glass Fiber/Polymer Composites*, p. 4, MO-SCI Corporation, Rolla, Missouri (June 30, 1993).
15. S. Breyer, *Breaking the Vicious Circle*, Harvard University Press, Cambridge, Massachusetts (1993).

10

The Future of Phosphate Fibers

The future of phosphate fibers should be very bright when the patents on the composition expire in 1999. In the author's mind there has never been a safety issue with phosphate fibers or any other inorganic phosphate that does not contain a toxic cationic or anionic function. Phosphates such as sodium, potassium, calcium, magnesium, ammonium, and hydrogen have never caused a problem of any type with perhaps the exception of renal stones. In this case it may be more of a condition than the cause of a condition. Urine is known to be saturated with respect to calcium phosphates. Some threshold agent or other mechanism that is poorly understood keeps kidneys from becoming completely calcified even in the healthiest of persons.

10.1. Interest in Phosphate Fibers Continues

Since the decision was made to discontinue efforts to commercialize phosphate fibers, two groups have continued to show intense interest in phosphate fibers. It is more or less expected that people working with composites should remain interested. The second group is the medical–dental community. Several groups at hospitals and universities have been using the fibers for dental composites and orthopedic prosthetics. Although both fibers have been requested, the $[Ca(PO_3)_2]_n$ fibers have been of more interest in bone and pin uses. It is not difficult to understand why vertebrates convert these fibers to more or less normal bone structures, as the polyphosphates are converted to apatites.

10.2. Japanese Research and Publications

The Japanese are pursuing research in phosphate fiber science with great vigor, particularly Professor Abe and his associates at the University of Nagoya. Perhaps as good an approach to this subject as any is to review publications and patent literature from Japan during the last eight to ten years. Some of these publications are particularly noteworthy. One of the more significant developments is a method of making microfibers of the same composition as the $[Ca(PO_3)_2]_n$ fibers.[1–5] Although several attempts were made to grow microfibers of this composition in our laboratory, we were never successful. Some microfibers based upon calcium pyrophosphate ($Ca_2P_2O_7$) were prepared, but they were never of very good quality.

10.3. Potassium Titanate Replacement

I chose to grow $[Ca(PO_3)_2]_n$ fibers in a sodium ultraphosphate melt. My reason for doing so was to obtain an amorphous water-soluble phase in which the $[Ca(PO_3)_2]_n$ fibers could grow, while attempting to avoid the crystallization of the calcium ultraphosphate, $Ca_2P_6O_{17}$. The Japanese chose to grow the fibers from the two-component $CaO–P_2O_5$ system without adding the Na_2O to their systems. They did indeed grow $Ca_2P_6O_{17}$ crystals along with their $[Ca(PO_3)_2]_n$ fibers. As expected, the $Ca_2P_6O_{17}$ crystals were easily dissolved from the $[Ca(PO_3)_2]_n$ fibers. Their $[Ca(PO_3)_2]_n$ fibers are beautiful microfibers with aspect ratios of as much as 130 and diameters as low as 1 to 5 μm (Figure 10.1). These fibers have magnificent properties and should be ideal for composite reinforcements similar to the potassium titanates, but without the dangers of potassium titanates.

$[Ca(PO_3)_2]_n$ fibers grown in the manner described above were more difficult to extract than the fibers grown in the sodium ultraphosphate melts, but their superior microfiber properties are a result of this growing medium and the orientation of fibers by aligning the glass in which they grew. The fibers grown in the two-component systems were leached with either distilled water or 0.1 molar sodium hydroxide solutions. Extraction periods of up to seventy-two hours were required to release the fibers from the matrix in which they grew.

10.4. Japanese Patents

Some of the Japanese patents that have been issued give direction as to how work is progressing from an industrial viewpoint and should act as a pattern for work that will follow in other parts of the world. One of the more interesting approaches was to prepare what has been called conjugate fibers. A water-soluble organic polymer is mixed with a water-soluble phosphate. The mixed polymers are spun in a hardening solution containing a water-soluble calcium salt. The hardened fiber is then fired to burn away the organic polymer, leaving behind what is claimed to be very strong single fibers of calcium phosphate.[6]

Several patents were applied for dealing with phosphate fibers and bone implants[7–10] while others dealt with frictional materials for brakes, clutches, and sheets.[11–15] There was very little that was novel in these patents when compared to

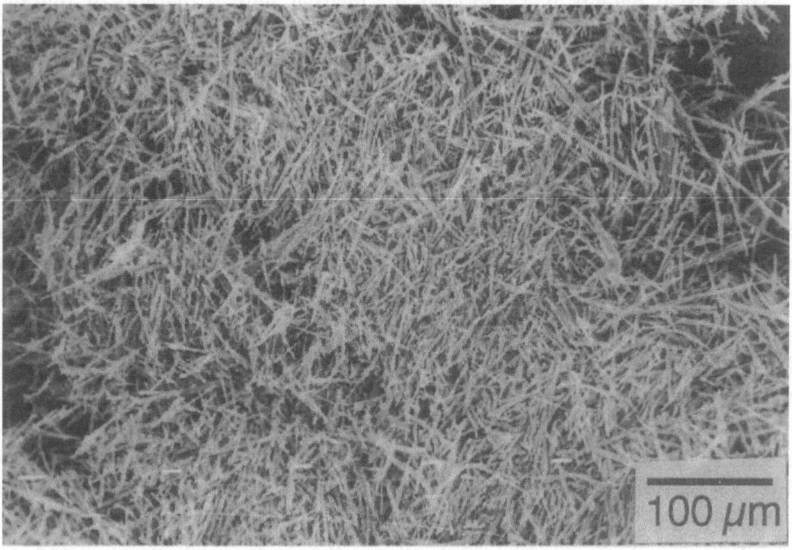

Figure 10.1 Professor Abe's $[Ca(PO_3)_2]_n$ phosphate fibers.

similar patents filed elsewhere in the world. In both cases the patents were based upon concepts that were more or less obvious to those skilled in the arts.

Two novel patents were applied for dealing with a method of making a zirconium phosphate fiber by a spinning technique[16] and another patent dealt with a method of making a smooth glass surface by admixing phosphate fibers in glass paper composition.[17] In the first example a soluble zirconium salt was used in much the same way that the soluble calcium salts were used in a spinning solution mentioned above.

Although past attempts to spin amorphous phosphate fibers have not been particularly promising, this area should not be neglected. There is a possibility that amorphous fibers can be spun and tempered, or perhaps even coated, to give them more resistance to hydrolytic degradation. Some new systems under study in our laboratory are showing some interesting possibilities, but it is still too early to predict the outcome. Although these are phosphates they are not related to the phosphate fibers that are featured in this book.

The work of Mr. Ngo initiated new fibers in our laboratory, but never progressed far enough to assess whether its true value is worthy of extensive new efforts. Of particular interest is Mr. Ngo's $[NaMg(PO_3)_3]_n$ fibers. These fibers are long and regular, silky, soft, and pliable. Of all the fibers found in the Phosphate Fibers Program, these were the most esthetically pleasing to both touch and sight. It was never learned why a double crystallization was required to grow these fibers, but it was probably nothing more complicated than driving most of the water of constitution from the lattice of the crystals. The first crystallization liberates most of the water of constitution while a second crystallization grows much longer chains, but this is conjecture.

Other fibers worthy of continued efforts are the long crystal of $[NaPO_3]_n$, Maddrell's salts; $[Na_3H(PO_3)_4]_n$ is similar to Maddrell's salt in many ways but is even more easily crystallized. $[LiCa(PO_3)_3]_n$, lithium calcium polyphosphates, as well as $[LiMg(PO_3)_3]_n$, lithium magnesium polyphosphates, were given only a cursory examination. None of these salts was investigated for its full potential as a useful fiber with a controlled rate of degradation in living systems. There is potential for both sodium and potassium Kurrol's salts in applications where film formers in a fibrous form could be helpful. Applications as bandages for bleeding wounds should be effective, since the alkali metal fibrous polyphosphates are good absorbers, yield solutions of viscous films, while precipitating proteins as polyphosphate coascervates. It will be remembered that this was one factor that was completely neglected in an ill-conceived rat study.

As often mentioned throughout this book, phosphate fibers have been used in foods for many years and as long as they were not considered substitutes for "asbestos" very little attention was given them in either their manufacture or their application. Many applications could be addressed that were only given a cursory glance when the project was in progress. When the decision was made to turn all attention to $[NaCa(PO_3)_3]_n$ by the Phosphate Fibers Project group, small efforts were continued by the Probes Group under my direction. When the main project ended, the Probes Group efforts were discontinued also because there was no reason to continue work on mineral fibers irrespective of their final usefulness. Attention was devoted to areas of more interest to our company.

10.5. Convert Natural "Asbestos" to a Biodegradable Fiber?

Converting "asbestos" into a safe product was discussed in detail in Chapter 5 but should be mentioned again as a future project. Several approaches could be successful in preparing "safe" chrysotile. Even admixes of polyphosphate fibers and chrysotile might very well cause the chrysotile to decompose as the phosphate is attacked. The degradation of the polyphosphates generates a mildly acidic condition and the phosphates are excellent sequestering agents of Mg^{+2}, and this action should accelerate the degradation of the asbestiform molecules.

Abe's Caution

It is evident that if the approach to making natural "asbestos" biodegradable should be followed, Abe's work with amorphous silicate–phosphate fibers cannot be ignored.[18,19] It is unlikely that fibers in which degradable linkages are inserted would be hazardous in any way unless they were used as implants. It seems ridiculous in most instances to implant materials into animals for study, unless it is intended that these substances are to be *used* as implants. There is no way this author could ever imagine how *large hunks* of any substance could enter the pleural cavity of a human by normal contact with the substance. Nor do I have any faith that the results of such tests mean anything concerning the safety of a substance, either natural or artificial.

10.6. New Chemistry and Physics

There is almost no limit to the new chemistry that could be researched to prepare new crystalline phosphate fibers with properties as varied as the compositions studied. It is expected that the best way to prospect for new fibers is to explore phase regions where the $M_2O–P_2O_5$ ratio is unity or less. M_2O in this case may be mixtures of metal oxides. M is one cationic equivalent by definition, but the cations may have charges of C^{+1}, C^{+2}, C^{+3}, etc. The sum of the equivalents of metal oxides should be less than or equal to the moles of P_2O_5. It is likely that the total composition should not stray too far away from a $M_2O–P_2O_5$ ratio near unity if the polyphosphate structures are desired as the framework on which the fiber is built. This is not intended as a statement of a theoretical limit and useful fibers may be found in other sections of a diagram. It is similar to stating that Manhattan is not prime area in which to hunt elephants unless there is a circus in town.

When working with M_2O–P_2O_5 near unity one must be on constant alert for the formation of ring metaphosphates rather than chain polyphosphates. I have promoted the concept that fibrous crystals are more likely to be grown from long slender molecules, but it was noted that metal fibers can be obtained which are not based upon slender molecules. As mentioned in many different ways throughout this study, the formation of chains or rings may be as fickle as seeds, composition, and cooling rates. It will be recalled that in the sodium system there are several sodium trimetaphosphate crystal forms, two or more sodium polyphosphate crystals as different as Maddrell's salt and Kurrol's salt, and these may be contained in a Graham's salt matrix. This is not limiting new studies but, on the contrary, adds to the very great range of interesting systems and properties remaining to be researched.

I have attempted to teach that there is no such item in the universe as *old chemistry. Old chemistry* breeds in the muddled thoughts of business types who are heard to say from time to time when referring to phosphates, "We have done research on phosphates. It is *old chemistry.*" In fact, we are woefully ignorant of phosphate chemistry and we have hardly scanned the flyleaf of this fascinating and useful science. Phosphate chemistry remains an industrial science because of its usefulness, but fortunately more universities and institutions are again doing new and exciting research. Universities in Japan, France, England, Russia, and even the United States are recognizing this research area as vital. Well-meaning, but sadly misdirected individuals have all but exterminated phosphorus chemistry. Detergent phosphates were the primary financial basis for most phosphorus research in both universities and industry. New uses for phosphorus are beginning to supplant detergents as an economic basis from which to derive funding. Phosphate fibers are a notable example.

Despite the fact that silicate–phosphate glasses showed some unexpected toxic influences, there are many other mixed anionic systems worthy of study. The borate–phosphate systems have been studied to a small degree. Some work has been done on sulfate–phosphate systems. Thilo's group studied the arsenate–phosphate and stibnate–phosphate chemistries in great detail, but very little applied work was done. This was probably a result of the mild toxicity of these systems.

Molecular-weight studies and chain-length determinations of both the $[NaCa(PO_3)_3]_n$ and $[Ca(PO_3)_2]_n$ fibers should be studied. It is known that their chain lengths as very long, but no formal studies were completed. Estimates based upon samples of fiber dissolved with aqueous solutions of disodium ethylene diamine tetraacetic acid should give a satisfactory starting point. Light scattering, intrinsic viscosity, centrifugation, or similar techniques should yield estimates that will probably be less than the crystalline chain lengths, because some hydrolytic degradation is certain to occur while dissolving and measuring the samples.

Very small degrees of degradation will surely drastically shorten chains that are extremely long before they are prepared for analysis. Even proximate analyses could be employed as indicators while using the equation introduced early in this study. In this case the relation between average chain length and M_2O–P_2O_5 ratio is used to calculated chain lengths. Water removal will be the major obstacle in this determination. Free water will act to lower the calculated chain lengths even if it is not a part of the phosphate molecules.

10.7. Deficiencies of Phosphate Fibers

Very little attention has been directed toward the deficiencies of phosphate fibers. There are many areas that need improvement. Fiber uniformity is one of the primary needs for $[NaCa(PO_3)_3]_n$ fibers, while decreased irritation and ease of preparation of the long $[Ca(PO_3)_2]_n$ fibers require more attention. The length and aspect ratio of $[NaCa(PO_3)_3]_n$ needs additional attention. A more satisfactory means of milling the blocky crystals to fibers is still needed. I proposed that milling should be a two-step operation with the samples first being crushed and then milled. This should give increased strength and increased fiber lengths.

10.8. The Need for an Unquestioned Mineral Fiber Remains

There remains a great need for a mineral fiber that is unquestioned as safe by responsible members of the regulatory agencies and the media.[20] This need is as much from a perceived danger of "asbestos" as it is from a real threat. As long as the citizen is told of a threat by persons believed to be reliable, the citizen has no other choice than to treat the threat as reality. The flame-fanners continue to support the concept that no fiber on this planet is safe, particularly "asbestos." It is unfortunate that the promoters of the hazards cause it to be almost impossible to offer any type of replacement, because replacements are all precondemned even before they become a reality.

Fireproof clothing will surely be manufactured from phosphate fibers in years to come. The advantages are too obvious and the means of preparing these yarns are not difficult to perfect. Some new techniques will be required, but there is no reason why it cannot be done.

Phosphate fibers were not commercialized because other options were judged to be better long-range risks for the future of our stockholders, irrespective of the public's health. Slow progress will always prevail when developing safe substances for our society in an environment of overstated hazards. Monsanto company management chose to risk the company's future on biological and pharmaceutical endeavors to cure the public's ills, rather than producing substances to prevent their ills. Who is to say whether or not this was right or wrong? Whether it was right or wrong is immaterial. Would *you* sponsor a safe replacement for "asbestos" knowing that lawsuits, sponsored by the very same "asbestos" flame-fanners, were certain to follow instantaneously? Under the circumstances perhaps the Monsanto management had no options.

Year in and year out the Monsanto management has been among the best in the world. They used their best judgement, at the time, and although the decision was not popular within the company, it must be respected. Hopefully, this respect, sometimes submerged by chiding, has surfaced throughout this book. I have spent a lifetime with Monsanto and I cannot believe that any other life would have been more rewarding. Monsanto makes every effort to be a good neighbor and the Monsanto management does care for you and your welfare! It is always the first consideration before any new project is undertaken. It can become very frustrating for those of us who try to bring you new and better products.

10.9. References

1. T. Kasuga and Y. Abe, *J. Inorg. Phos. Chem.* **2**, 17 (1992).
2. T. Kasuga, A. Ichino, and Y. Abe, *J. Ceram. Soc. Japan* **100**, 1088 (1992).
3. T. Kasuga, H. Hosono, and Y. Abe, *Phosphorus Sulfur Silicon* **76**, 247 (1993).
4. Y. Abe, M. Hosoe, T. Kasuga, H. Ishikawa, N. Shinkai, Y. Suzuki, and J. Nakayama, *J. Am. Ceram. Soc.* **65**, 189 (1982).
5 Y. Abe, T. Kasuga, H. Hosono, and K. de Groot, *J. Am. Ceram. Soc.* **67**, 142 (1984).
6. Y. Abe, H. Hosono, T. Kasuga, and M. Nagase, *Phosphorus Sulfur Silicon* **76**, 243 (1993).
7. Y. B. Peng and J. B. Koen, *Bioabsorbable Glass Fiber/Polymer Composites*, p. 4, MO-SCI Corporation, Rolla, Missouri (June 30, 1993).
8. T. Yutaka, Japanese patent 04-24117 [JP 424117 A] (August 28, 1992).
9. T. Hideo *et al.*, Japanese patent 60-017118 [JP 60017118 A] (July 6, 1983).
10. T. Hideo *et al.*, Japanese patent 62-069823 [JP 62069823 A] (September 12, 1986).
11. S. Yasusaburo, Japanese patent 62-072938 [JP 62072938 A] (September 6, 1985).
12. S. Tetsuya, Japanese patent 63-053326 [JP 63053326] (August 21, 1986).
13. O. Hiroyasu *et al.*, Japanese patent 01-045437 [JP 1045437 A] (August 19, 1987).
14. M. M. Kuratsuchifuiirudo *et al.*, Japanese patent 61-213226 [JP 61213226 A] (March 13, 1986).
15. M. M. Kuratsuchifuiirudo and J. A. Hinkebein, Japanese patent 61-160499 [JP 61160499 A] (December 23, 1985).
16. K. Ryohei *et al.*, Japanese patent 55-003337 [JP 55003337 A] (June 22, 1978).
17. E. Masatoshi, Japanese patent 63-303876 [JP 63303876 A] (December 12, 1978).
18. T. Kasuga, H. Hosono, and Y. Abe, *Phosphorus, Sulfur Silicon* **76**, 507 (1993).
19. T. Kasuga, A. Ichino, and Y. Abe, *Nippon Seramikkusu Kyokai Gakujutsu Ronbunshi* **100**(Aug.), 1088 (1992).
20. W. A. Onderick, *Dealing With Asbestos, Plant Engineering,* March, 102 (1994).

Index

MIX
Papier aus verantwortungsvollen Quellen
Paper from responsible sources
FSC® C105338
FSC
www.fsc.org

If you have any concerns about our products,
you can contact us on
ProductSafety@springernature.com

In case Publisher is established outside the EU,
the EU authorized representative is:
Springer Nature Customer Service Center GmbH
Europaplatz 3, 69115 Heidelberg, Germany

Printed by Libri Plureos GmbH
in Hamburg, Germany